Muhammer Bulat

Molecular cluster cations of carbon monoxide and carbon dioxide

Muhammer Bulat

Molecular cluster cations of carbon monoxide and carbon dioxide

development of a reflectron time-of-flight mass spectrometer for cluster surface impact studies

Südwestdeutscher Verlag für Hochschulschriften

Impressum/Imprint (nur für Deutschland/only for Germany)
Bibliografische Information der Deutschen Nationalbibliothek: Die Deutsche Nationalbibliothek verzeichnet diese Publikation in der Deutschen Nationalbibliografie; detaillierte bibliografische Daten sind im Internet über http://dnb.d-nb.de abrufbar.

Alle in diesem Buch genannten Marken und Produktnamen unterliegen warenzeichen-, marken- oder patentrechtlichem Schutz bzw. sind Warenzeichen oder eingetragene Warenzeichen der jeweiligen Inhaber. Die Wiedergabe von Marken, Produktnamen, Gebrauchsnamen, Handelsnamen, Warenbezeichnungen u.s.w. in diesem Werk berechtigt auch ohne besondere Kennzeichnung nicht zu der Annahme, dass solche Namen im Sinne der Warenzeichen- und Markenschutzgesetzgebung als frei zu betrachten wären und daher von jedermann benutzt werden dürften.

Coverbild: www.ingimage.com

Verlag: Südwestdeutscher Verlag für Hochschulschriften GmbH & Co. KG
Dudweiler Landstr. 99, 66123 Saarbrücken, Deutschland
Telefon +49 681 37 20 271-1, Telefax +49 681 37 20 271-0
Email: info@svh-verlag.de

Approved by: Berlin, Humboldt Universität zu Berlin, Diss., 2010

Herstellung in Deutschland:
Schaltungsdienst Lange o.H.G., Berlin
Books on Demand GmbH, Norderstedt
Reha GmbH, Saarbrücken
Amazon Distribution GmbH, Leipzig
ISBN: 978-3-8381-2920-4

Imprint (only for USA, GB)
Bibliographic information published by the Deutsche Nationalbibliothek: The Deutsche Nationalbibliothek lists this publication in the Deutsche Nationalbibliografie; detailed bibliographic data are available in the Internet at http://dnb.d-nb.de.

Any brand names and product names mentioned in this book are subject to trademark, brand or patent protection and are trademarks or registered trademarks of their respective holders. The use of brand names, product names, common names, trade names, product descriptions etc. even without a particular marking in this works is in no way to be construed to mean that such names may be regarded as unrestricted in respect of trademark and brand protection legislation and could thus be used by anyone.

Cover image: www.ingimage.com

Publisher: Südwestdeutscher Verlag für Hochschulschriften GmbH & Co. KG
Dudweiler Landstr. 99, 66123 Saarbrücken, Germany
Phone +49 681 37 20 271-1, Fax +49 681 37 20 271-0
Email: info@svh-verlag.de

Printed in the U.S.A.
Printed in the U.K. by (see last page)
ISBN: 978-3-8381-2920-4

Copyright © 2011 by the author and Südwestdeutscher Verlag für Hochschulschriften GmbH & Co. KG and licensors
All rights reserved. Saarbrücken 2011

Abstract

This thesis deals with the metastable decay and the surface scattering induced fragmentation in the hyperthermal energy range of relatively weakly bound molecular cluster cations. With $(CO)_n^+$ and $(CO_2)_n^+$ two related model systems were chosen for a systematic size dependent study. Surface impact experiments were carried out with stainless steel and SiO_2 covered Si(100) surfaces. Results were obtained by a new, compact reflectron time-of-flight mass spectrometer (Re-TOFMS). Additional to the experimental data we present in this work a detailed description of the instrumental design considerations, numerical optimization, ion optical simulations and construction. We discuss mass resolution and resolution limiting effects in Re-TOFMS. Hence each ion optical component like electron guns, accelerator, deflector, mass gate and reflectron are described in detail. Numerical optimization and ion optical simulations were used to develop a compact instrument with high resolving power and mass selection capability. Despite the compact dimensions with a total flight length of ≈ 1.5 m the developed instrument possesses a high mass resolution above $m/\Delta m = 3000$. Additionally it offers the possibility to perform mass separation of big cluster ions with sizes $n \leq 190$. As a result mass-selected cluster ions can be studied for metastable decay channels and for interactions with surfaces. As a model system small carbon dioxide cluster ions $(CO_2)_n^+$ with $n \leq 15$ were mass selected and collided with the stainless steel surface backplane of the reflectron collider. In that case the reflectron collider was utilized as an energy analyzer. Metastable decay channels and the origin of fragmentation products were determined by kinetic energy analysis. Comparable measurements with small carbon monoxide cluster ions $(CO)_n^+$ with $n \leq 40$ impacted on a stainless steel surface and SiO_2 covered Si(100) silicon surface were performed, too. For the cluster ions of both molecules no evidence for shattering was observed even for relatively high collision energies $E_i \geq 500$ eV. In the case of both cluster types metastable decay via unimolecular dissociation was observed for the electron ionization or impact heated parent cluster.

Keywords:
Molecular Clusters, Metastable Decay, Surface Impact,
Time-of-Flight Mass Spectrometer

Zusammenfassung

Diese Dissertation handelt vom metastabilen Zerfall und von der Oberflächenwechselwirkung im hyperthermalen Energiebereich relativ schwach gebundener molekularer $(CO)_n^+$ und $(CO_2)_n^+$ Clusterionen mit einer Edelstahloberfläche und einer mit der natürlichen SiO_2 Oxidschicht belassenen Si(100) Siliziumoberfläche. Im Rahmen dieser Arbeit wurde ein hierfür geeignetes spezielles Flugzeitmassenspektrometer entwickelt und aufgebaut. Entwurf, numerische Optimierung der Auflösung, ionenoptische Simulationen und Aufbau der jeweiligen Komponenten wie Elektronenquellen, Beschleuniger, Ablenkplatten, Massenfilter und Reflektron werden detailliert beschrieben. Das entwickelte Flugzeitmassenspektrometer besitzt mit einer kompakten Gesamtfluglänge von ≈ 1.5 m eine hohe Massenauflösung von $m/\Delta m \geq 3000$. Es bietet die Möglichkeit, eine Massentrennung von Clusterionen mit einer Größe von bis zu $n \leq 190$ vorzunehmen. Diese massenselektierten Clusterionen können daraufhin auf metastabilen Zerfall und auf ihre Wechselwirkung mit einer Oberfläche untersucht werden. Dazu wurden Kohlendioxid-Clusterionen $(CO_2)_n^+$ mit $n \leq 15$ massenselektiert und mit einer im Reflektron platzierten Edelstahloberfläche kollidiert. Hierbei wurde das Reflektron als Energieanalysator eingesetzt. Über die kinetische Energie der Eltern-Clusterionen und der Fragmentionen kann auf metastabile Zerfallskanäle und Herkunft der Fragmente geschlossen werden. Vergleichbare Messungen wurden auch mit kleinen Kohlenmonoxid-Clusterionen $(CO)_n^+$ mit $n \leq 40$ an einer Edelstahloberfläche und an einer Si(100)-Siliziumoberfläche vorgenommen. Für die Clusterionen der beiden Moleküle war auch für hohe Kollisionsenergien ($E_i \geq 500$ eV) kein kompletter Zerfall in Monomere nach der Oberflächenwechselwirkung nachweisbar. Aus den experimentellen Beobachtungen wurde für die metastabilen Elternclusterionen der beiden Moleküle geschlossen, dass diese sowohl bei der Anregung durch Elektronenstoßionisation als auch durch Oberflächenstoß durch das Abdampfen von Monomeren abkühlen.

Schlagwörter:
Molekülcluster, Metastabiler Zerfall, Oberflächenstoß,
Flugzeit-Massenspektrometer

To my family.

Rigid Body Sings[*]

Gin a body meet a body
    Flyin' through the air,
Gin a body hit a body,
    Will it fly? and where?
Ilka impact has its measure,
    Ne'er a ane hae I,
Yet a' the lads they measure me,
    Or, at least, they try.

Gin a body meet a body
    Altogether free,
How they travel afterwards
    We do not always see.
Ilka problem has its method
    By analytics high;
For me, I ken na ane o' them,
    But what the waur am I?

JAMES CLERK MAXWELL (1831–1879)

---

[*]dialect translation: "gin" = if... "ilka" = every... "ane" = one... "hae" = have... "a' " = all... "ken" = know... "waur" = worse...

L. Campbell. *The Life of James Clerk Maxwell, with a selection from his correspondence and occasional writings and a sketch of his contributions to science.* Macmillan, London, 1882.

Contents

1	**Introduction**	**1**
	1.1 Research Objectives	3
	1.2 Thesis Content	4
2	**Basic Principles**	**5**
	2.1 The Molecular Beam	5
	2.1.1 Beam Temperature and Velocity	7
	2.1.2 Cluster Generation	11
	2.1.3 Size Distribution	12
	2.2 TOF Mass Spectrometry	14
	2.2.1 Progress in TOFMS	14
	2.2.2 Basic TOF Principles	14
	2.2.3 Advanced TOF Principles	15
	2.2.4 Resolution	16
	2.2.5 Resolution Optimization	21
	2.2.6 Optimization Procedure	27
	2.3 Metastable Decay	29
	2.4 Cluster-Surface Interactions	31
3	**Experimental Setup**	**36**
	3.1 Assembly and Vacuum System	36
	3.2 Cluster-Ion Generation	38
	3.2.1 Pulsed Nozzle	38
	3.2.2 Electron Guns	38
	3.3 Detection Sytems	39
	3.3.1 Faraday Cup	39
	3.3.2 MCP-Detector	40
	3.4 Electronics	41
4	**Results and Discussion**	**42**
	4.1 TOFMS Optimization	42
	4.1.1 Numerical Optimization	42
	4.1.2 The Accelerator	49
	4.1.3 The Deflector	56
	4.1.4 The Mass Gate	65

		4.1.5	The Reflectron and Target Surface	70
	4.2	TOFMS Spectra		80
		4.2.1	Mass Calibration	80
		4.2.2	Linear TOFMS Mass Resolution	83
		4.2.3	Reflectron TOFMS Mass Resolution	84
		4.2.4	Mass Separation	85
		4.2.5	Cluster Size and Intensity	85
	4.3	Metastable Decay and Surface Impact		97
		4.3.1	Impact of $(CO_2)_n^+$ on Stainless Steel Surface	97
		4.3.2	Impact of $(CO)_n^+$ on Stainless Steel Surface	113
		4.3.3	Impact of $(CO)_n^+$ on SiO_2 covered Si(100) Surface	129

5 Summary and Outlook — 142
   5.1 Summary — 142
   5.2 Outlook — 147

Bibliography — 148

List of Figures — 171

A TOFMS — 176
   A.1 Pictures — 176
      A.1.1 TOFMS Accelerator — 176
      A.1.2 TOFMS Deflector — 177
      A.1.3 TOFMS Mass Gate — 177
      A.1.4 TOFMS Reflectron — 178
      A.1.5 Retarding Field Energy Analyzer — 178
   A.2 SIMION Geometry Files — 179
      A.2.1 Three Stage TOFMS Accelerator — 179
      A.2.2 TOFMS Deflector — 181
      A.2.3 TOFMS Reflectron — 182

B Electron-Guns — 184
   B.1 Nozzle Mounted Electron-Gun — 184
   B.2 Flange Mounted Electron-Gun — 185

Chapter 1
Introduction

> *"There are two ways to live your life. One is as though nothing is a miracle. The other is as though everything is a miracle."*
> ALBERT EINSTEIN (1879–1955)

In the present time many researchers are concerned with the scientific evidence on climate change. Accordingly the interest in greenhouse gases such as carbon dioxide showed a steep increase in the last decades. Since the industrial revolution the concentration of carbon dioxide (CO_2) and carbon monoxide (CO) in the atmosphere increased rapidly by the use of fossil fuels.

Apart from that carbon monoxide is the most abundant interstellar molecule[1] next to hydrogen [1]. Hence under interstellar or atmospheric conditions such particles can be ionized by radiation and undergo chemical reactions. Here the most important aspect with respect to ion molecule reactions is that carbon monoxide and carbon dioxide cations are possible precursors of amino acids. In that sense the most exciting challenge is the detection of amino acids e. g. glycine (NH_2CH_2COOH) in the interstellar media which are the basic building blocks required for the development of life. Despite theoretical predictions no successful detection of amino acids in interstellar media has been reported in literature to date [2].

On the other hand an important process for the climate is the formation of atmospheric aerosol particles (several nanometers in diameter), which are also known as clusters when small [3]. These particles are formed by nucleation and subsequent growth of e. g. ionized germs. Clusters cover the intermediate state between the single atom or molecule and their corresponding macroscopic bulk matter phase. Within this intermediate state the chemical and physical properties of the atom or molecule evolve to the chemical and physical properties of the macroscopic bulk matter phase. These properties depend on the cluster size, the number of constituent molecules forming the cluster. Consequently size dependent studies on clusters require sophisticated mass spectrometric techniques for the size selection and detection of the sample. Largest changes of the chemical and physical properties were observed within the range of the smallest cluster sizes beginning

---

[1] typically a factor of about 10^{-4} or more compared to hydrogen

with the dimer. Generally the evolution of "nanoscopic" (few-body) to macroscopic (many-body) physical and chemical properties shows no linearity for small cluster sizes [4–8]. Thus Castleman et al. [9] describe clusters as "superatoms" which extend the periodic table to the third dimension. According to Castleman et al. clusters are "superatoms" which provide an unprecedented ability to design novel nanostructured materials.

However, while a lot is known about the properties of single atoms or molecules and their corresponding macroscopic bulk phases much less is known about the intermediate state covered by the clusters. Hence over the last three decades the interest in cluster research has increased rapidly. Most of the studies focused on the size dependent chemical and physical properties of the clusters (e. g. electronic, magnetic, optical, structural and reactive) [6; 10–12]. One of these interesting properties is the binding energy which changes in a nonlinear way in dependence of the molecular unit, size and charge of the cluster. According to Mähnert et al. [13], with 1.80 eV the carbon monoxide dimer shows one of the largest reported binding energy values for an ionized van der Waals dimer. Hiraoka et al. [14] reported a steep size dependent decrease of the binding energy value for small $(CO)_n^+$ cluster ions with $3 \leq n \leq 18$. The binding energy (total, average and per unit) of the cluster is of great importance in determining the stability and structure of a cluster [15]. Therefore experimental effort is required to investigate the evolution of such properties in dependence of the cluster size. One possible method is interacting the cluster with a well defined surface. Accordingly, sophisticated instruments were introduced to generate size selected clusters of various atoms or molecules and to collide these size selected clusters with a surface at well defined collision energies [16; 17]. Depending on the collision energy, surface and "sample" cluster many different processes were observed. The processes are non-dissociative scattering, impact induced dissociation, fission and evaporation, cleavage, mechanical bond splitting, energy dissipation, intracluster reactions, transient bond formation with the surface, electron transfer, ion-pair formation, cluster anion electron emission, secondary-electron and -ion emission from a surface,... [17]. One of the interesting processes is the surface collision induced dissociation (SID) of clusters. In that case the collision energy is partially converted to internal energy of the cluster which can exceed the binding energy of the cluster. Such high internal energies can lead to evaporation, metastable fragmentation, cleavage and "shattering" of the excited cluster. Therefore the question arises whether the strongly cluster size dependent binding energy values of small CO clusters do influence surface interaction, particularly surface impact induced fragmentation and impact induced "shattering". Furthermore, it would be interesting to compare CO with another well known sample molecule. In that case with CO_2 a comparable model system with quite different chemical and physical properties is available. Besides, both molecules are strongly related to each other by chemistry e. g. catalysis [5; 8; 18; 19].

However, despite the popularity of the CO_2 and CO molecule as model systems to the best of my knowledge no cluster size dependent studies about the surface interaction of CO_2 and CO molecular cluster ions exist.

1.1 Research Objectives

The aim of this work was to study the size dependent cluster surface interaction and metastable decay of relatively weakly bound molecular cluster cations. Therefore, for these studies with CO and CO_2 two interesting and simple model systems were chosen. Accordingly such a study requires a suitable experimental setup to generate, ionize, size select, and impact these cluster ions on a well defined surface. Hence, prior to the experimental work it is essential to design, simulate, develop, setup and test such a custom made device. Besides these technical challenges additionally it is necessary to solve many experimental challenges e.g.:
Neutral molecular clusters can be generated by molecular beam expansion. Molecular beam sources generate clusters with a broad size distribution. Therefore the expansion pressure, temperature of the sample gas and the ratio of a possible seed gas must be optimized to acquire control over the cluster size distribution. However, generally only ionic species can be utilized and detected by mass spectrometric analysis. Consequently suitable ion sources must be designed, optimized and tested for the maximization of the ion signal intensities. Apart from that the mass spectrometric device should allow size selection of the desired cluster size and the adjustment of the desired collision energy prior to the surface impact. Equally important is the fact that parent ions, metastable daughter ions and surface impact induced product ions could be detected and distinguished from each other with the same device. Consequently for these studies an optimized instrument with high resolving power, transmission and detection sensitivity is required.

1.2 Thesis Content

Chapter 2 gives a short overview of the basic principles of supersonic molecular beam expansion, cluster formation and cluster size distribution. Also included is a section about time-of-flight mass spectrometry principles with the focus on resolution and resolution optimization. This chapter ends with short introductions about metastable decay and cluster surface interactions.

Chapter 3 includes a brief description of the experimental setup.

Chapter 4 begins with the results about the numerical optimization of the time-of-flight mass spectrometer. Accordingly in this section design criteria are discussed by SIMION simulations for the accelerator geometry, deflector geometry and reflectron geometry. Afterwards, the resulting mass resolution and mass selection performance of the developed device are discussed by means of mass spectra. This section is followed by a section about the ionization parameters and their influence on the cluster size distribution. The last section of this chapter deals with the metastable decay and the cluster surface impact results. Accordingly results of the metastable decay and surface impact with the stainless steel surface of carbon dioxide and carbon monoxide cluster cations are discussed for different cluster sizes. The last section in this chapter contains the results of the metastable decay and surface impact of carbon monoxide cluster cations with the SiO_2 covered $Si(100)$ surface.

Chapter 5 includes a short summary of the thesis results and an outlook for future work.

Chapter 2

Basic Principles: Molecular Beams, Time-of-Flight Mass Spectrometry, Metastable Decay and Cluster-Surface Interactions

> *"This result is too beautiful to be false; it is more important to have beauty in one's equations than to have them fit experiment."*
> PAUL ADRIEN MAURICE DIRAC (1902–1984)

2.1 The Molecular Beam

Since the innovative experiments of Stern and Gerlach [20] molecular beams advanced to a versatile tool in physics, chemistry and engineering. Extensive studies with supersonic molecular beams have been performed since then to probe surface properties [21–25], to transport molecules into the gas phase [26], to generate atomic and molecular clusters [27; 28] and for many other processes. The molecular beam expansion driven transfer of molecules into the gas phase is also coupled with an efficient cooling of molecules below 1 K [29–31] which is of vast importance for spectroscopy [32–36]. The molecular beam emerges from the expansion of a probe gas in a stagnation vessel (stagnation pressure p_0 and temperature T_0) through an orifice or nozzle with diameter d into vacuum (or lower pressure p_b). Regarding a high pressure p_0 in the reservoir, the mean free path λ of the probe particles (atoms or molecules) in the stagnation vessel is several orders of magnitude smaller than the diameter of the expansion orifice. The mean free path λ is defined by

$$\lambda = \frac{k_B T_0}{\sqrt{2} p_0 \sigma} \qquad (2.1)$$

with $k_B = 1.38065 \times 10^{-23}$ J/K the Boltzmann constant and σ the collision cross section defined by $\sigma = \pi D^2$ with D the effective collision diameter of the particles [37] ($D_{(CO)} = 3.8$ Å and $D_{(CO_2)} = 3.68$ Å [38]). The ratio λ/d defines the emerging

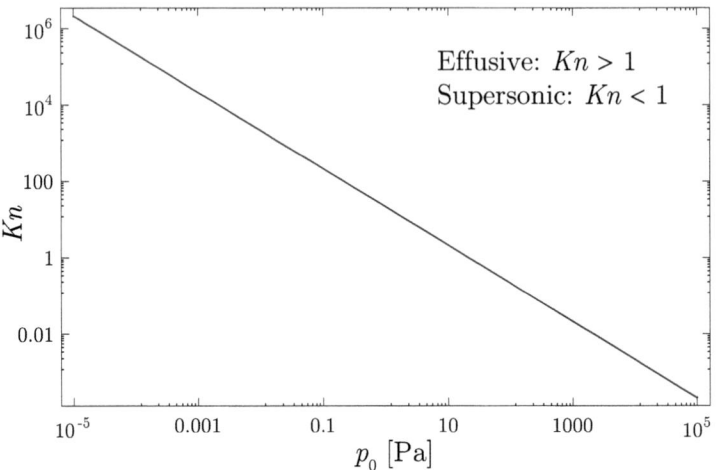

Figure 2.1 Logarithmic plot of the Knudsen number Kn in dependence of the stagnation pressure p_0. Calculated for fixed values: temperature $T_0 = 300$ K, $D_{(CO)} = 3.8$ Å and $d = 300$ μm. With increasing stagnation pressure p_0 (assuming an expansion into the vacuum), Kn reaches very fast values with $Kn < 1$ which pertain to the supersonic expansion regime.

flow properties and is known as the Knudsen number Kn [37; 39–41]:

$$Kn = \frac{\lambda}{d} \qquad (2.2)$$

The Knudsen number permits to distinguish different flow regimes and describes the degree of rarefaction in the beam. For molecular beam expansion into vacuum the regime with $Kn < 1$ is of interest. In that case the mean free path λ is much smaller than the orifice diameter d (see drop in Kn with increasing stagnation pressure p_0 in figure 2.1). The particles in the stagnation vessel are pushed through the orifice or nozzle by the pressure gradient and enter the vacuum chamber. Hence they lose their randomized velocities which they possessed in the stagnation vessel and obtain one main propagation direction and velocity. Thus a great part of the total energy of the gas in the reservoir is converted to kinetic energy. Hence, the mean velocity of the particles increases. During the expansion many collisions between the particles flowing through the orifice take place. Due to the collisions a quasi equalization of the velocities is established narrowing the velocity distribution in the beam. In the case that the energy exchange of the particles with the orifice or nozzle and the background gas is negligible the expansion can be regarded as an adiabatic expansion. In the sense that this expansion is a reversible process it is also isentropic.

2.1.1 Beam Temperature and Velocity

From the assumption of an ideal gas and an one-dimensional adiabatic expansion process the mean beam velocity $\langle u_\| \rangle$ on the axis can be deduced easily [42–44]. The approach is done by the fundamental law of energy conservation during the expansion. The stagnation enthalpy H_0 in the reservoir is given by $H_0 = H + \frac{1}{2}m\langle u_0 \rangle^2$ (according to the first law of thermodynamics). During the expansion of the gas the temperature and the enthalpy decreases whereby the mean velocity $\langle u_\| \rangle$ increases. For an ideal gas the change in enthalpy can be written in the form $dH = c_p dT$, with c_p the temperature independent heat capacity at constant pressure ($c_p = (\partial H/\partial T)_P$). The stagnation enthalpy is converted partially into kinetic energy $\frac{1}{2}m\langle u_\| \rangle^2$ of the directed mass flow and a rest enthalpy H [45; 46]:

$$H_0 + \text{const.} = c_p T_0 = H + \text{const.} + \frac{1}{2}m\langle u_\| \rangle^2 = c_p T_\| + \frac{1}{2}m\langle u_\| \rangle^2 \qquad (2.3)$$

Regarding the energy conservation (2.3) and the assumption of an ideal gas, the mean flow velocity $\langle u_\| \rangle$ on the axis is directly correlated to the decrease in temperature $T_\|$ in the form:

$$\langle u_\| \rangle^2 = \frac{2}{m}\int_{T_\|}^{T_0} c_p dT = \frac{2c_p}{m}(T_0 - T_\|) \qquad (2.4)$$

The maximum mean flow velocity $\langle u_\| \rangle_{\text{max}}$ of the beam is reached when $T_\|$ in equation (2.4) drops to negligibly low values ($T_\| \ll T_0$). In this case the maximum possible mean flow velocity $\langle u_\| \rangle_{\text{max}}$ can be expressed by the following equation,

$$\langle u_\| \rangle_{\text{max}} = \sqrt{\frac{2c_p}{m}T_0} = \sqrt{\frac{2k_B}{m}\frac{\gamma}{\gamma-1}T_0}, \qquad (2.5)$$

where γ is the ratio of the specific heats $\gamma = c_p/c_v$ at constant pressure and volume. In the case of an "ideal" gas c_p is equal to $\frac{5}{2}k_B$. In equation (2.5) it is assumed that the randomized translational velocities of the particles in the stagnation chamber are converted to a directed flow in one main translational direction in consequence of the supersonic expansion. Rearrangement of equation (2.3) using $k_B = c_p - c_v$ results in an equation for the final parallel temperature $T_\|$ of the beam [47]:

$$T_\| = T_0 \left[1 + \frac{1}{2}(\gamma-1)M^2\right]^{-1} \qquad (2.6)$$

In the equation (2.6) above M is the the local Mach number. With γ being temperature independent (ideal gas) $T_\parallel$ will depend only on the local Mach number M and T_0. However, for real gases such as CO and CO_2 the ratio of the specific heats γ depends on temperature and pressure and shows large values in close vicinity of the critical point [31]. The local Mach number M is defined by the ratio of the stream velocity $u_\parallel$ to the local speed of sound c (in the case of an ideal gas: $c = \sqrt{\gamma k_B T_\parallel / m}$). It is evident from the equation (2.6) that the local temperature $T_\parallel$ in the beam decreases with increasing Mach number M and vice versa. Regarding the isentropic expansion of an ideal gas, this means that M will increase drastically along the expansion path. This behavior originates from the decrease of the speed of sound c which decreases as $\sqrt{T_\parallel}$ resulting in large Mach numbers. During expansion the increasing Mach number exceeds $M \geq 1$ which gives reason for labeling the expansion as a *supersonic molecular beam*.

In the case of a supersonic beam the velocities are not much higher than in a normal effusive[1] beam as the increase in M is caused by the decrease of the local sound velocity c [39; 43; 47; 48]. During the expansion of the gas through a nozzle, the particle density and pressure perpendicular to the beam propagation direction fall off dramatically and reach pressure values below p_b the background pressure present in the expansion chamber. At this point no further beam expansion is possible in these directions which can be described as reaching the boundary conditions (see figure 2.2). The gas jet leaving the nozzle cannot sense the boundary conditions since information is transported by the beam only at the speed of sound. At the free jet boundaries this leads to pressure values below p_b resulting in succeeding recompression by shock waves (*barrel shock*) described as overexpansion [37; 42; 49]. These shock waves form regions of temperature, pressure, density and velocity gradients. The shape and characteristic features of a supersonic jet expansion are depicted schematically in figure 2.2. In molecular beam experiments generally the core of the expansion is extracted by a skimmer [48; 52–54] for further use. For sufficiently low background pressure p_b the core of the beam is nearly not influenced by the boundary conditions, in that sense the flow in this region is isentropic and also referred to as *zone of silence*. A Mach shock wave oblique to the beam propagation direction is also formed and called Mach disk. Up to the Mach disk the flow reaches its terminal Mach number M_T which depends on Kn, γ and a particle size specific prefactor (collision effectiveness). Assuming an ideal gas the terminal Mach number for Ar can be estimated by [39; 50]

$$M_T = 1.17 \times Kn^{(1-\gamma)/\gamma}, \tag{2.7}$$

---

[1]Effusive beams are formed e. g. by effusion of a gas from an oven or other sources through an orifice into a vacuum chamber. In the case of effusive beams the mean free path λ is much larger than the diameter of the expansion orifice d (no collisions when passing through the orifice).

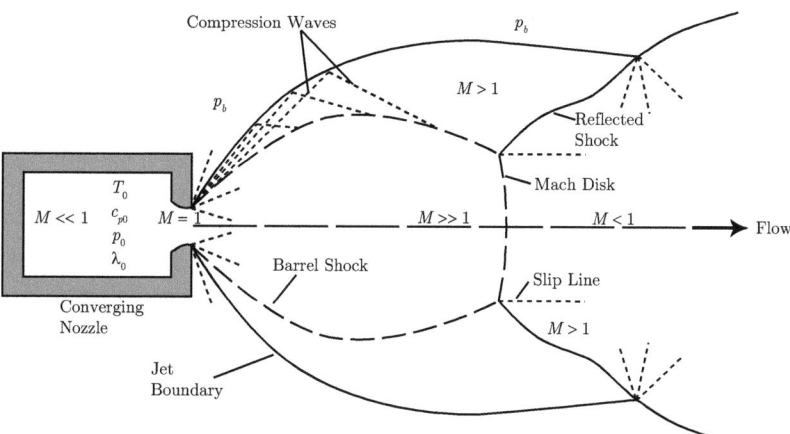

Figure 2.2 The schematic shape of a supersonic molecular beam expansion with its different regions after [50]. The shape of a free jet expansion and the rotational temperatures in a supersonic jet of CO_2 were visualized by Raman mapping [51].

where 1.17 is an experimentally obtained value [39]. In the case of e.g. carbon monoxide a terminal Mach number $M_T = 16.7$ was reported (for $T_0 = 218 \pm 5$ K, $T_\infty = 3.9 \pm 0.1$ K (terminal temperature) and $\gamma = 7/5$) [55] (for CO_2 and other gases see [56; 57]). The terminal Mach number is reached when no more collisions take place in the flow. Hence at long distance from the nozzle the Mach number and also the temperature reach asymptotically terminal values [47]. The location of the Mach disk x_m is measured in nozzle diameters d and can be calculated by the following empirical expression [37; 54]:

$$\frac{x_m}{d} = 0.65\sqrt{\frac{p_0}{p_b}} \qquad (2.8)$$

Experimental results show that equation (2.8) can be used for various gases including monatomic and diatomic molecules [58]. The velocity distribution of the particles in the beam is of Maxwellian nature. It is best described by the superposition of two Maxwellian velocity components, the perpendicular ($\perp$) and parallel ($\parallel$) one. Such a distribution is called an ellipsoidal normalized velocity distribution [39; 49; 59; 60],

$$f(\mathbf{v})d\mathbf{v} = n\sqrt{\frac{m}{2\pi k_B T_\parallel}} \left(\frac{m}{2\pi k_B T_\perp}\right) \times \exp\left(-\frac{m(v_\parallel - u)^2}{2k_B T_\parallel} - \frac{mv_\perp^2}{2k_B T_\perp}\right) d\mathbf{v}, \qquad (2.9)$$

which according to Toennies et al. [59] reduces to the usual Maxwellian distribution function when $T_\parallel = T_\perp$. The beam temperature is here defined by the velocity spread $\Delta v_\parallel$ of the particles underlying a Maxwellian distribution. Hence the velocity spread of a Maxwell-Boltzmann distribution $\Delta v_\parallel$ is used for the temperature derivation. One meets such velocity distributions in the literature in various forms [31; 46; 60–64]. The most common equation is seen below:

$$f(v_\parallel)dv_\parallel = cv_\parallel^2 \exp\left[-\left(\frac{v_\parallel - \langle u_\parallel \rangle}{\Delta v_\parallel}\right)^2\right] dv_\parallel \qquad (2.10)$$

Here c denotes a scaling factor for the centerline beam intensity. Experimental determination of beam temperatures is done by the measurement of molecular beam velocity distributions. One obtains a more convenient form of equation (2.10) for application in experiments with the Jacobian transformation from the velocity domain to the time domain. In case of a flux sensitive detector also a conversion is needed. The resulting distribution of flight times for a flux sensitive detector is then given by

$$f(t)dt = c\frac{L^3}{t^4} \exp\left[-\left(\frac{L/t - \langle u_\parallel \rangle}{\Delta v_\parallel}\right)^2\right] dt, \qquad (2.11)$$

with L being the total flight distance between the beam source and the detector. For a density sensitive detector the term L^3/t^4 in equation (2.11) is then replaced by the term L^2/t^3 [28; 65].

Hence evaluation of the parallel beam velocity spread $\Delta v_\parallel$ is carried out by fitting molecular beam time of flight distributions with equation (2.11). The resulting velocity spread $\Delta v_\parallel$ of the distribution is directly related to the translational temperature by [31; 44; 49; 66]:

$$\Delta v_\parallel = \sqrt{\frac{2k_B T_\parallel}{m}} \qquad (2.12)$$

The measured velocity distributions are then usually characterized by the speed ratio S which is defined in terms of the velocity $\langle u_\parallel \rangle$ and the parallel beam velocity spread $\Delta v_\parallel$ of the distribution, with

$$S = \frac{\langle u_\parallel \rangle}{\Delta v_\parallel} = \frac{\langle u_\parallel \rangle}{\sqrt{\frac{2k_B T_\parallel}{m}}}. \qquad (2.13)$$

Higher speed ratios S correspond to lower translational temperatures $T_\parallel$. For helium already speed ratios greater than 1000 were reported corresponding to temperatures lower than 1 mK [29]. For supersonic jet expansion of CO and CO_2

efficient cooling in close vicinity of their critical points was observed with speed ratios above 100 corresponding to translational temperatures $T_\|$ below 0.1 K [31].

2.1.2 Generation of Clusters in Molecular Beams

Cluster type	Prototypical cases	Binding forces	Average binding energy (eV)
Van der Waals clusters	(Rare gases)$_n$ (N$_2$)$_n$ (CO$_2$)$_n$	Dispersive plus weak electrostatic	≤ 0.3
Molecular clusters	(I$_2$)$_n$, (organics)$_n$	Dispersive, electrostatic (weak valence)	~ 0.3 to 1
Hydrogen-bonded clusters	(H$_2$O)$_n$, (NH$_3$)$_n$	H-bonding, electrostatic	~ 0.3 to 0.5
Ionic clusters	(NaCl)$_n$	Ionic bonding	~ 2 to 4
Valence clusters	C$_n$, S$_8$	Conventional chemical bonds	~ 1 to 4
Metallic clusters	Na$_n$, Cu$_n$	Metallic bonds	~ 0.5 to 3

Table 2.1 Classification of binding properties of different cluster systems, after Märk [15]

The term cluster is described in the Oxford dictionary[2] as a collection of "things" of the same kind; a bunch. In cluster physics and chemistry the word "things" stands for atoms or molecules (see table 2.1). Clusters bridge the gap between atoms (or molecules) and the condensed bulk phase. So the cluster size N ranges from the dimer with $N = 2$ to e. g. $N = 10^5$ and up to microcrystals or microdroplets [27; 67]. Cluster formation in molecular beams was first reported in 1956 by Becker et al. [68]. They observed an increased beam intensity and higher beam velocity which they accounted for by condensation in the beam. The formation of clusters in a supersonic beam is a complicated process and until today no complete theoretical description exists [69]. Macroscopically the condensation process can be explained by the supersaturation of an expanding gas. The onset of condensation depends then in general on the source conditions (stagnation pressure and temperature) [70]. Roughly this process can be treated as a gas-liquid phase transition. The adiabate of the supersonic expansion crosses the vapor pressure curve and leads to high supersaturation in the molecular beam. At this point cluster formation sets in with releasing concurrently condensation

---
[2]"cluster, n." The Oxford English Dictionary, 2nd ed. 1989, OED Online, Oxford University Press, 2000, http://dictionary.oed.com/cgi/entry/50042182

heat [43; 67]. Microscopically the formation of clusters can be described by two- and three-body collisions. During the adiabatic expansion of a gas (see subsection 2.1.1) many collisions take place in the beam. The local translational temperature in the beam reaches very fast very low values. In this case cluster formation is evident also for slightly bound systems interacting via van der Waals forces (see table 2.1)[71]. Cluster growth starts with the aggregation of two free particles to dimers [43; 44; 49; 72; 73]:

$$A + A \longrightarrow A_2^* \qquad (2.14)$$

Due to the release of binding energy the dimer can be formed in an excited state (2.14). In that sense an additional collision partner is needed for the conservation of momentum and energy (therefore three-body collisions)[3]:

$$A_2^* + M \longrightarrow A_2 + M^* \qquad (2.15)$$

The excess energy is absorbed by the third collision partner, in general a monomer (2.15). As a result in time-of-flight experiments an increase in monomer velocities was reported which was accounted by cluster formation in the beam [68; 74; 75]. Bigger clusters emerge from dimers which act as condensation nuclei for further growth by collisions with monomers or other clusters. Reaching the free molecular region where no collisions take place, the growth of the clusters stagnates. Remaining condensation heat is transported away from the cluster by the evaporation of monomers or bigger fragments [76; 77]. In the case of high background gas pressure above 10^{-1} mbar destruction of clusters due to heating up by scattering with residual gas cannot be neglected [78]. In addition to the mentioned condensation from the gas phase very big clusters can also be formed by droplet formation e.g. by fragmentation of a liquid droplet in liquid jets [66; 79–83].

2.1.3 Cluster Size Distribution

Small clusters may show strong nonlinear size dependent properties [9], therefore a prior size selection for the investigation of these properties is required. However, size selection can reduce the signal intensity whereas most experiments require (or benefit from) high intensities of the sample. Thus, it is of vast importance to control, maximize or predict the size or at least the average cluster size $\langle N \rangle$ produced in the beam. Due to the lack of a rigorous theory this attempt led to the semiempirical scaling laws for cluster formation introduced by Hagena [67; 70; 84–86]. The sizes of clusters produced in a jet expansion basically depend on the stagnation conditions. By variation of the source parameters T_0, p_0, nozzle diameter and shape a wide range of cluster sizes can be formed. The onset of clustering usually is described by the reduced dimensionless scaling parameter Γ^*

---

[3]Molecular dimers can also be formed by two-body collisions where the excess energy is transfered to vibrational or rotational modes.

[84] referred to as the Hagena parameter given by [87; 88]:

$$\Gamma^* = k \frac{(d/\tan\alpha)^{0.85}}{T_0^{2.29}} p_0 \qquad (2.16)$$

where d is the nozzle diameter in μm, α is the expansion half angle and k a constant related to bond formation ($k = 1650$ for Ar and $k = 3660$ for CO_2, for other gases see [88]). For the scaling parameter $\Gamma^* < 200$ no clustering in experiments was observed. The transition from a flow without condensation to a flow with cluster formation was observed for $200 < \Gamma^* < 1000$. Bigger clusters with sizes exceeding 100 formed by massive condensation were formed for $\Gamma^* > 1000$. Contrary to the generated average cluster size of neutrals $\langle N \rangle$ the cluster size distribution of ionized species can be measured easily e. g. with retarding field energy analysis [67; 70; 89–91]. However in this case a prior ionization of the clusters with a suited method e. g. electron ionization is necessary (see also subsection 3.2.2). Due to the narrow velocity distribution in the supersonic jets, all cluster sizes possess nearly the same velocity (except velocity slip). Hence their kinetic energy is mainly affected by their mass. Assuming singly charged particles the potential of the retarding field can be used to derive the mass by the following relation:

$$\frac{1}{2} m u^2 = e U_f \qquad (2.17)$$

where e is the elementary charge and U_f the energy filter retarding potential. Nevertheless it must be kept in mind that the interaction of fast electrons (usually a few ten eV up to a few hundred eV) or intense laser pulses [92] with the cluster during ionization heats up the cluster. This can result in fragmentation and evaporative dissociation of monomers due to the increased temperature of the system [93; 94]. Therefore the cluster size distribution of neutral clusters differs from that of cluster ions [44; 87; 95]. The size distribution of neutral clusters can be measured by the utilization of optical measurement setups [83; 88; 96; 97] or with crossed beam scattering techniques [98–100]. As will be discussed later on (see subsection 4.2.5) the cluster size distributions observed by mass spectrometry are of log normal nature [101; 102]. However, some cluster sizes can be observed which dominate their neighbors in the mass spectra by higher intensities and are referred to as "magic numbers". The appearance of "magic numbers" in mass spectra was reported for the first time for rare gas xenon clusters by Echt et al. in the year 1981 [103]. Reports on the observation of magic numbers for other systems followed soon [11; 104–108]. The preferential formation of clusters with sizes of $N = 13, 19, 55, \ldots$ can be attributed to the geometrical structure of these clusters. Due to energy reasons the icosahedral closed shell structures of magic sized clusters are more stable than the open shell neighboring cluster sizes.

2.2 Time-of-Flight Mass Spectrometry

2.2.1 Progress in Time-of-Flight Mass Spectrometry

In the year 1946 Stephens [109] proposed to build a mass spectrometer based on the flight time dispersion between accelerated ions of different mass to charge ratio. Two years later the first device based on this principle was constructed by Cameron and Eggers [110]. But the newly introduced so called "ion velocitron" suffered from its poor resolution. Problems and solutions for improving resolution of Time-of-Flight Mass Spectrometers (TOFMS) were discussed in the comprehensive publication of Wiley and McLaren [111]. Since 1955 designs of most TOFMS are based on the publication of Wiley and McLaren. With the development of pulsed lasers in the mid-1960's TOFMS obtained a well suited ionization source. Additionally, lasers made it possible to probe surface compositions with TOFMS. However, up to 1972 there was no further significant improvement in TOFMS techniques regarding resolving power. The instrumental innovation of Mamyrin et al. [112] was the crucial step in enhancing resolving performance of TOFMS. With the innovation of the ion mirror by Mamyrin et al. and the implementation of new ionization and desorption sources like secondary laser ionization [113], electrospray ionization [114] and laser and plasma desorption, TOFMS evolved to a wide spread method. Beside this up to that time advances in electronics and detectors also pushed the application of TOFMS on and improved further the performance. The relevance of TOFMS increased two decades ago with the development of matrix assisted laser desorption (MALDI) by Hillenkamp et al. [115; 116]. Today MALDI-TOFMS is indispensable for the analysis of large biomolecules with masses of several thousands of atomic mass units (amu). On the other side new miniature laser ablation TOFMS are constructed for in situ planetary exploration with acceptable resolution [117], and show one important field of application of modern TOFMS apparatuses. A good overview over the field of time-of-flight mass spectrometry can be found in various review articles [118–122]. Due to its modular buildup, it's easy to upgrade TOFMS instruments or combine it with other devices to hybrid systems like quadrupole-TOFMS. A brief introduction to the field of quadrupole-TOFMS can be found in the review of Guilhaus et al. [123] and Chernushevich et al. [124].

2.2.2 Basic Time-of-Flight Principles

Time-of-flight mass spectrometry is a separation in time technique. It is principally based on the conversion of electric field energy to kinetic energy deducible by elementary Newtonian mechanics [125]. A resting charged particle with the charge $q = ne$ (with n an integer) in an electric field is forced to move along the streamlines of the field. The particle moves from a position with a higher potential value U_h to a position with a lower potential value U_l. By this movement the initially resting particle gains kinetic energy defined by the potential difference between these positions. In that sense for a motion limited in one dimension and without other fields the following equation is valid:

$$\frac{1}{2}mv^2 = q(U_h - U_l) = q\Delta U \quad (2.18)$$

The equation (2.18) is comparable with equation (2.17). In the case of (2.17) a moving particle is decelerated by a retarding field and in the case of (2.18) a particle is accelerated by an extraction field. Rearrangement of the equation (2.18) delivers the basic relation between the mass of the particle and the velocity after the acceleration by the potential difference.

$$v = \sqrt{\frac{2q\Delta U}{m}} = \sqrt{z\Delta U}, \quad \text{with} \quad z = \frac{2q}{m} \quad (2.19)$$

If the potential or the charge of the particle does not change in (2.19), it is obvious that the velocity and thus the time of flight of the particle depend only on the mass m. Due to conservation of energy all particles with different masses will gain the same amount of kinetic energy by traversing the potential difference. This means that lighter particles will have higher velocities than heavy particles. The substitution of the velocity v in (2.19) by the traversed distance L divided by time t will relate the time of flight directly to the mass of the particle given by:

$$t = \frac{L}{v} = L\sqrt{\frac{1}{z\Delta U}} \quad (2.20)$$

2.2.3 Advanced Time-of-Flight Principles

Basic understanding of ion-optics is required to understand the principles of a TOFMS apparatus. However, to design such an apparatus it would be helpful to have knowledge of advanced ion-optics [126–128]. In this section the basic principles given in the subsection (2.2.2) are expanded to the main time-of-flight principles. A detailed and extended description of the fundamental time-of-flight theory can also be found in [129]. In general a basic linear TOFMS consists of two main regions, an acceleration region and a field free drift region with a detector located on its end. Before ionization the sample substance is normally available in the gas phase or can be desorbed from a surface. Ionization occurs before entering the acceleration region or inside the acceleration region. The most common ionization methods are laser-ionization [130] or electron impact ionization [131; 132]. In the Wiley McLaren configuration the accelerator consists of two acceleration stages defined by three electrodes enclosed by wire meshes (see figure 2.3). The first electrode is the repeller and the third electrode the grounding mesh. The potentials applied to the repeller and middle mesh define

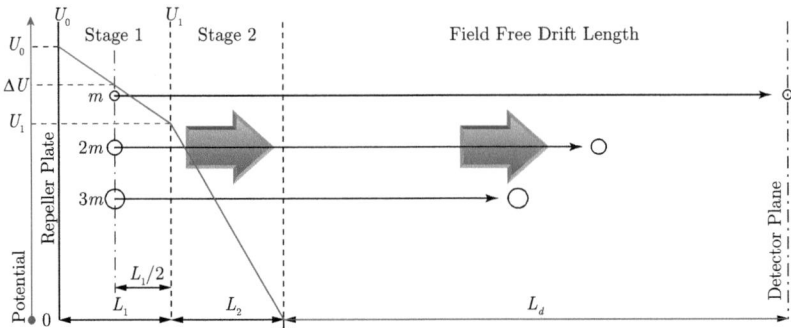

Figure 2.3 The schematic principle of a two stage Wiley-McLaren type linear TOFMS in conjunction with a potential diagram. Ions with different masses m start in the middle of the first acceleration stage. Every ion gains a kinetic energy defined by the mass to charge ratio and the potential difference ΔU (see equation 2.19). The lightest ion will reach the detector at the shortest time.

the acceleration voltage. The ions enter the first stage or are generated in it and are accelerated with the application of a voltage pulse to the repeller and middle mesh. At this time ideally every ion in the accelerator gains the same kinetic energy by the electric field gradient between repeller and the grounded plate. If the applied voltage pulse is long enough every ion will enter the field free drift region with the same kinetic energy. Therefore, the ion velocities differ and depend only on the mass to charge ratio $1/z$ of the ions. Ions with lower masses will reach the detector at the end of the field free region faster than ions with higher masses. The flight times of the ions will be proportional to the square root of their $1/z$ ratio (see eq. 2.19). If the used ionization method delivers singly charged ions with $q = e$, the registered flight time of the ions at the detector can be related to the mass of the ion. An additional requirement to do this is that the detector is fast enough to indicate the arrival of each ion and sensitive enough to record the ions. For very slow ions generally big organic molecules special technical arrangements must be taken for detection [133]. Besides this problem an overview of the physical and technical problems and their particular solutions in time-of-flight mass spectrometry is summarized by Guilhaus et al. [134].

2.2.4 Resolution and Resolution Improvement

Resolution or resolving power of the instrument defines its capability to discriminate between two neighboring ions in the mass spectra of nearly the same mass to charge ratio [135]. With the idealized view described above (2.2.2), every ion with the same mass to charge ratio will gain the same kinetic energy by the acceleration

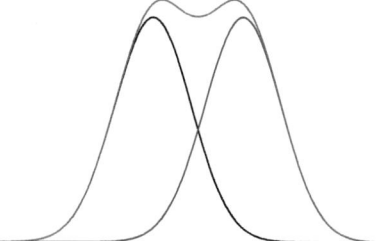

Figure 2.4 The FWHM definition of mass resolution. Two adjacent and overlapping Gaussian peaks (black and blue) appear in the mass spectra as the convolution of both peaks (red). The two Gaussian peaks are distinguishable when the valley is just discernible [135].

field and will arrive at the same time on the detector. So in a mass-spectrum we would see these ions as a very sharp peak with a width defined by the response-time of the detector. This would admit a very high mass resolution only limited by the asymptotic behavior of the square root function (2.20) for extremely high masses. However resolution limiting effects will prohibit this due to flight time differences for the ions with the same mass to charge ratio. Resolution limiting effects result in deviations in flight times and can be regarded as "flight-time errors". The main errors in flight times are caused by the initial kinetic energy distribution σ_v and the initial starting positions σ_x of the ions in the accelerator [136; 137]. So some ions will need a "turn-around-time" in the case of an inverse initial velocity according to the acceleration direction. Thus the calculation and optimization of the resolution requires an error analysis of peak broadening effects. By the consideration of the two main resolution limiting factors, 1. the spatial distribution of ion starting positions σ_x and 2. the initial velocity (energy) distribution σ_v the variance of the iso-mass peak σ_t can be approximated by the following equation [138],

$$\sigma_T^2 = \left(\frac{\partial T}{\partial x_0}\right)^2 \sigma_x^2 + \left(\frac{\partial T}{\partial v_0}\right)^2 \sigma_v^2, \qquad (2.21)$$

here x_0 is the initial ion position, v_0 the initial ion velocity and T the total ion flight time ($T(x_0, v_0)$). In the case that the ions are generated in the acceleration region, a factor for temporal distributions of ion formation times $\sigma_{t_0}^2$ must be added in equation (2.21) to the right hand side [139]. Equation (2.21) is just valid for infinitesimal changes in initial ion position or initial energy. Assuming that each factor has a Gaussian contribution a more valuable approach for the peak shape can be made based on probability theory [138; 140]. Other contributions to the error in flight-times are due to inhomogeneous fields in the acceleration region or ion-optics, unstable power supplies, deflection of the ions at grids, inaccuracy of the detection system and so on. The real TOFMS mass peak recorded in the mass spectra is much broader than expected for an ideal spectrometer. In this case peaks in the high mass region can overlap for two neighboring masses making it impossible to distinguish between these two masses. With the assumption that the mass peak shapes are Gaussian in nature the resolution can be defined as the discriminability criterion between two neighboring mass peaks [135]. The resolution of the apparatus is defined by the sharpness of the detected mass peak (m) respectively by its width at half peak-maximum (Δm_{FWHM} or Δt_{FWHM}, see figure 2.4). In this case the resolution r is given by:

$$r = \frac{\Delta m_{\mathrm{FWHM}}}{m} = \frac{2\Delta t_{\mathrm{FWHM}}}{t} \quad (2.22)$$

Here the resolution is typically expressed in ppm. In the literature, mostly the resolving power R [135] with

$$R = \frac{1}{r} = \frac{t}{2\Delta t_{\mathrm{FWHM}}} \quad (2.23)$$

is also referred to as resolution. In the following the second definition for resolution R (2.23) is used. To maximize the resolution of a TOFMS it is possible to reduce the resolution limiting effects by careful design. To limit the effect of the "turn-around-time" a narrow beam source, which delivers a supersonic molecular beam with low thermal kinetic energy, should be used [138] (see also 2.1.1). Additional high extraction fields will also limit the effect of the initial kinetic energy distribution. With higher extraction voltages the ions will reach the detector faster and the time dispersion will be narrower. However this is limited due to electronics, mesh-flexing and sparkovers of high voltages. With orthogonal extraction [134] of the ions it is also possible to reduce the initial velocity in extraction direction. Using a long field free flight region will extend the time-of-flight and simultaneously the resolution. But here the limit is given by geometric requirements and by the vacuum-technique required for the evacuation of big volumes. Hence most TOFMS have short drift tubes and possess under normal conditions poor resolution. However the resolution can be significantly enhanced by adding an ion-mirror (reflectron, see page 25) [112] for the compensation of the initial velocity. Besides the improvement of the instrumental apparatus with better power supplies or faster detectors, several methods are described in the literature for achieving better mass resolution in time-of-flight mass spectrometry. To compensate the initial ion starting distribution space focusing is applicable. The compensation of the initial velocity and spatial distribution of the ions can be achieved by time-dependent ion-extraction. This was first proposed by Wiley and McLaren and called "time-lag energy focusing" [111]. A similar method was introduced by Browder et al. in the form of impulse-field focusing theory

[141]. They describe the improvement of resolution of a TOFMS apparatus by application of a very high and shortly pulsed ion extraction field followed by the conventional pulsed extraction field. A quite different focusing method known as dynamic-field focusing was proposed by Yefchak et al. [142]. This model is based on the dynamic post source acceleration of ions at the space-focus plane. This means a second acceleration of ions which arrive at the space focus plane. The space focus plane is the position in the field free drift region, where the ions of the same mass have the lowest time-of-flight differences (smallest error in time-of-flight distribution, thus highest resolution). In general a detector or a mass gate (see subsection 4.1.4) is placed at the space focus plane. The methods of dynamic ion extraction described above for improving energy resolution are important for sources with high initial ion velocities and can improve resolution for a narrow mass range. These methods are well suited for example for MALDI-TOFMS [116; 143]. The major limitation on resolution and mass accuracy in MALDI-TOFMS originates in the relatively broad distribution of initial velocities of ions produced by the laser desorption process [129; 144]. However TOFMS apparatuses using cold molecular beams with very low velocity distributions are not affected by this problem like MALDI apparatuses. In that case a different approach is needed. The main resolution limiting effects in molecular beam source TOFMS are the combination of initial velocity distribution and initial starting positions of the ions [139]. Besides the time dependent extraction of ions another possibility is to optimize the acceleration potentials or length of the acceleration stages to maximize the achievable resolution. This process is called focusing. In contrast to light optics, focusing in TOFMS means a focusing in flight-times (e.g. moving the focus plane to the detection plane). It involves the minimization of time-of-flight distributions of ions with the same mass but different starting conditions (velocity and position). Minimization of time-of-flight errors resulting from initial starting positions of the ions is called space focusing (see figure 2.5). Analogously the minimization of time-of-flight errors caused by the initial velocity distribution of the ions is called energy focusing. First work in this direction was done by Wiley and McLaren who introduced a two stage TOFMS with first order "space focusing". In this case the acceleration voltages are calculated in the way to compensate the flight-time difference of iso-masses caused by the distribution of initial starting-positions. This method is just valid for small changes in initial starting-positions. Wiley and McLaren obtained the parameters for first order space focusing by an analytical treatment of the total flight time function [111]. The condition for first order space focusing is that the first derivative of TOF with respect to the initial starting position vanishes:

$$\frac{\partial T(x_0, v_0)}{\partial x_0} = 0 \qquad (2.24)$$

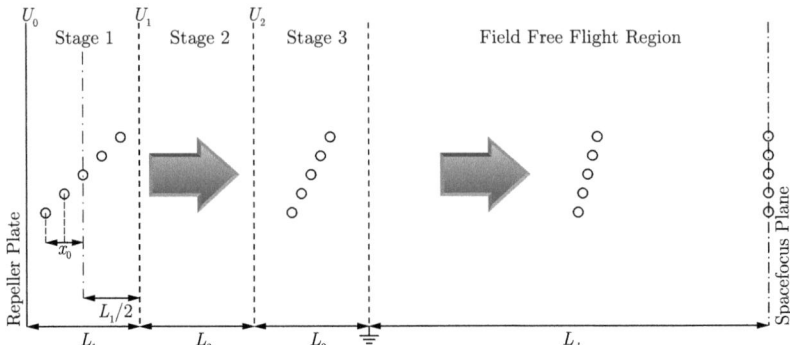

Figure 2.5 Depicted is the principle of space focusing in a three stage TOFMS. Ions start at different positions x_0 in the first acceleration stage. Space focusing is achieved by adjusting the lengths L_i or the potentials U_i. At the space focus plane all ions with the same mass will have the lowest time-of-flight deviation.

They set the first derivative of the function to zero and extracted the parameters for the space focusing condition (see equation (2.24)). Analogously the condition for first order energy (velocity) focusing is that the first derivative of TOF with respect to the initial starting velocity will vanish:

$$\frac{\partial T(x_0, v_0)}{\partial v_0} = 0 \qquad (2.25)$$

So it depends on the source conditions if space focusing or energy focusing will improve resolution. First order focusing condition is achieved when the first derivative vanishes. For higher order focusing besides the first order the higher order derivatives also must vanish. This can be treated analytically by expressing the deviations in TOF as a Taylor series and setting the derivatives equal to zero [145; 146]. However, it is not possible to achieve energy focusing and space focusing simultaneously analytically [146]. Therefore it depends on the source conditions for the TOFMS which kind of focusing will deliver the better result. Space focusing will be well-suited e.g. for narrow velocity distributions produced by cold orthogonally extracted molecular beam source. On the other hand in the case of MALDI-TOFMS where the ions are generated by laser desorption from a probe surface the ions possess high thermal energies (velocities) in one direction. In that case energy focusing will deliver a better result than space focusing. According to Reddish et al. second and higher order focusing requires more parameters like additional acceleration stages and lengths for the optimization process [145]. They show in their theoretical treatment that it would be easier and more reliable to use three or more acceleration stages than two to fulfill the second order space focusing condition. It was shown earlier that second order space focusing with just two acceleration stages [147–149] delivers superior resolution in contrast to the Wiley McLaren configuration (see figure 2.3). The idea of using more acceleration stages than the two of the original Wiley McLaren configuration is not new [150] and was implemented by Even and Dick [146; 151] with success. For higher order focusing closed analytical solutions of (2.24) and (2.25) are not available. Therefore the numerical optimization of resolution was introduced as an alternative approach [151].

2.2.5 Numerical Optimization of Resolution

Numerical optimization allows the calculation of the parameter sets for optimal focusing for a given system. On the other hand it is possible to simulate and optimize the design before setting up a TOFMS device. For an accelerator in Wiley-McLaren configuration (see figure 2.3) first order space focusing can be obtained easily. With an additional effort second order focusing will also be possible. Extending the two stage system to a three stage system (see figure 2.5) will allow second order space focusing and an improved resolution compared to the Wiley-McLaren configuration. Therefore a comparison of the performance of a three stage system with a two stage system is of interest. In the following a detailed description of the numerical optimization of resolution of a multistage TOFMS device will be given. Based on this method the TOFMS apparatus introduced in this work was designed and constructed. To simplify the calculation and the complexity of the theory the problem was reduced to one dimensional motion of the ions. The numerical optimization is done as follows:

- Definition of the geometrical parameters (fixed parameters and parameters which will be optimized),

- formation of an ion group (10001) arranged in a line symmetrically around $L_1/2$ (spatial distribution),

- calculation of the starting velocity v_0 for each ion in the group (according the velocity distribution),

- calculation of the time-of-flight for each ion,

- calculation of the variance in flight-times for a given L_d,

- calculation of the resolution in dependence of the TOF variance and the spatial distribution probability of each ion,

- the optimization routine searches in the parameter space for the parameter set which delivers the highest resolution (loop).

So the main problem in numerical optimization is to minimize the variance in flight times with changing the initial parameters like the stage potentials or the geometrical parameters L_i. This can be done with minimization algorithms which are implemented in most calculus software. In Mathematica [152] this can be done by the function "NMinimize". The comparable function in Matlab [153] is referred to as "fminsearch". For our computations we used the function "optim" in the program "R" [154]. The advantages of the program R are that it is open source and powerful enough to be used on older computer systems (e.g. a Pentium II system). In general most minimizing functions in these programs are based on "Nelder-Mead", "quasi-Newton" and "conjugate-gradient" algorithms. Best results were obtained by the use of the "L-BFGS-B" method which was introduced by Byrd et al. [155] and allows box constraints (an improved version of the "quasi-Newton method").

The Accelerator For the optimization process at first the calculation of the flight times in dependence of the accelerator parameters (potentials and lengths) is required. We consider an orthogonally extracted supersonic molecular beam with a narrow transversal velocity distribution. The extraction is done pulsed, so the beginning of the extraction pulse defines the time zero t_0. The ions start in a symmetric line in the first stage around the half length $L_1/2$ of the first stage (see figure 2.5). Depending on their starting positions x_0, the ions possess different potential energies. The potential energy is defined by the potential energy difference of the two meshes which form the first acceleration stage with respect to the starting position inside this stage. It is assumed that the meshes are ideally parallel to each other. Then, of course, the potential will decrease linearly from the first mesh to the second one. In that case the potential energy E_{pot} of an ion starting in the position x_0 can be described by the following equation,

$$E_{\text{pot}} = q\Delta U = q(U_1 - U_0)\left(\frac{x_0}{L_1} - 1\right), \qquad (2.26)$$

where U_0 is the potential of the first grid (also referred to as repeller), U_1 the potential of the second grid and L_1 the length of the first acceleration region (with $U_0 > U_1$). The formula (2.26) describes the potential energy gain of an ion when it leaves the first stage depending on its starting position. To calculate the time of flight of the ion an approach based on the law of energy conservation similar to eq. (2.18) can be formulated. In that case the ion velocity obtained in the first acceleration stage can be related to the ion starting position x_0 (potential energy) and the ion initial velocity $\pm v_0$ (initial kinetic energy) with:

$$\frac{1}{2}mv_1^2 = q(U_1 - U_0)\left(\frac{x_0}{L_1} - 1\right) \pm \frac{1}{2}mv_0^2 \qquad (2.27)$$

Equation (2.27) can be solved for the velocity v_1 and we obtain an equation similar to equation (2.19):

$$v_1(x_0, v_0) = \sqrt{(U_1 - U_0)\left(\frac{x_0}{L_1} - 1\right)z \pm v_0^2} \qquad (2.28)$$

However, v_0 is added if its orientation is in extraction direction and is subtracted vice versa. In the same way we can calculate the velocity the ion will obtain after leaving the second and the third stage:

$$v_2(x_0, v_0, v_1) = \sqrt{v_1(x_0, v_0)^2 + (U_1 - U_2)z} \qquad (2.29)$$

$$v_3(x_0, v_0, v_1, v_2) = \sqrt{v_1(x_0, v_0)^2 + v_2(x_0, v_0, v_1)^2 + U_2 z} \qquad (2.30)$$

With basic Newtonian mechanics for accelerated motion the corresponding flight times spent by the ions in each acceleration stage can be calculated by the following equations:

$$t_1(v_0, v_1) = \frac{2(L_1 - x_0)}{v_1(x_0, v_0) \pm v_0} \qquad (2.31)$$

$$t_2(v_1, v_2) = \frac{2L_2}{v_1(x_0, v_0) + v_2(x_0, v_0, v_1)} \qquad (2.32)$$

$$t_3(v_2, v_3) = \frac{2L_3}{v_2(x_0, v_0, v_1) + v_3(x_0, v_0, v_1, v_2)} \qquad (2.33)$$

where L_2 and L_3 are the length of the second and the third acceleration stage. After leaving the acceleration stages the ion possesses the velocity v_3 and travels with this velocity to the detector which is placed in a distance L_d from the accelerator. The ion will require the time t_d for arriving on the detector given by:

$$t_d(v_3) = \frac{L_d}{v_3(x_0, v_0, v_1, v_2)} \qquad (2.34)$$

The whole time-of-flight t_{all} spent by the ion beginning with the "time zero" and ending with arriving at the detector is the sum of all flight times spent in each region.

$$t_{\text{all}} = \sum_{i=1} t_i \qquad (2.35)$$

Now equation (2.35) allows us to calculate the total time-of-flight of an ion depending on its initial position and initial velocity (for the linear TOFMS configuration, for the reflectron TOFMS configuration see page 25). Thus we can start any number of ions for calculating their standard deviation $\sigma_{t_{\text{all}}}$ in flight times, with:

$$\sigma_{t_{\text{all}}}^2 = \sum_{i=1}^n \left[t_{\text{all}}(x_0, v_0, i) - \overline{t_{\text{all}}}(x_0, v_0, i) \right]^2 p(x_0, i) \quad (2.36)$$

where $p(x_0, i)$ is the probability that the ion i will start at the initial position x_0. For simplicity one can use an equipartition function and the probability is then $p(x_0, i) = 1/n$ (with n defining the number of ions). But for realistic environment simulations like a skimmed molecular beams one must consider the source properties what will be the subject of the following part. With $\overline{t_{\text{all}}}(x_0, v_0, i)$ being the average time-of-flight and the Gaussian nature of the distribution, the time deviation $\sigma_{t_{\text{all}}}$ also underlies a "Gaussian-distribution". Consequently the resolution resulting from this deviation can be expressed by:

$$R = \frac{\overline{t_{\text{all}}}(x_0, v_0, i)}{2\sqrt{\ln 4} \cdot \sigma_{t_{\text{all}}}} \quad (2.37)$$

The inverse of R in equation (2.37) can be used for minimization in the optimization process. Thus the resolution R will be maximized by finding the parameters which solve the minimization problem of the reciprocal value of equation (2.37). The first question that arises is how the initial velocity v_0 of the ions can be estimated? This can be done by a simple assumption: An ion with no transversal velocity component ($v_0 = 0$) will start in the middle of the first acceleration stage at $x_0 = L_1/2$. So an ion which starts at the right side of $L_1/2$ must have a transversal velocity component that is positive. Analogously an ion that starts on the left side of $L_1/2$ must have a transversal velocity component that is negative. If we consider the time zero and assume that all the ions obtain the same axial velocity so the offset from $L_1/2$ defines v_0 by:

$$v_0 = \frac{(x_0 - L_1/2)}{t_s} \quad (2.38)$$

where t_s is the time-of-flight from the skimmer to the acceleration region. If the valve is placed in a distance of 100 mm and assuming for simplicity that the expansion velocity of the gas is 1000 m/s, the ions will need 10^{-4} s to reach the acceleration region. With respect to the time t_s at time zero an ion that is starting with an offset of 4 mm from $L_1/2$ must have a transversal velocity component of ± 40 m/s. Besides the initial velocity distribution an additional question arises regarding the spatial distribution. Under realistic experimental conditions there is not an equipartition distribution of the ions in the accelerator. Hence the spatial distribution is more complex than $p(x_0, i) = 1/n$. It can be assumed that the cluster ions produced by the supersonic nozzle source obey a Maxwell-Boltzmann velocity distribution $f(v_0)$:

$$f(v_0) = \sqrt{\frac{m}{2\pi k_B T_\perp}} \exp\left(-\frac{v_0^2 m}{2 k_B T_\perp}\right) \quad (2.39)$$

The spatial distribution $\phi(x)$ of the ions is obtained by the convolution of the space dependent transformation $f(x)$ of $f(v_0)$ with a function $h(x)$:

$$\phi(x) = f(x) * h(x) \quad (2.40)$$

In equation (2.40) $h(x)$ describes the shape of the used skimmer and $f(x)$ the space dependent transformation of equation (2.39) which is given by:

$$f(x) = \sqrt{\frac{m}{2\pi k_B T_\perp}} \frac{1}{t_s} \exp\left(-\frac{m}{2 k_B T_\perp} \frac{x^2}{t_s^2}\right) \quad (2.41)$$

For a conical skimmer with a circular shape and radius r one can use as $h(x)$ the following function:

$$h(x) = \pm\sqrt{r^2 - x^2} \quad (2.42)$$

For avoiding complexity it is more convenient to use a step-function for $h(x)$ representing a skimmer with rectangular shape. By estimating a beam temperature $T_\perp$, the spatial distribution of the ions can be calculated for a given skimmer geometry. For supersonic beam sources $T_\perp$ can be narrower than 1 K defined by geometrical circumstances. For our calculations we assume for the molecular beam a temperature of $T_\perp = 1$ K. The written program uses this temperature to calculate the spatial distribution with equation (2.40). The convolution is done by the function "convolve" implemented in R. Thus the obtained spatial distribution is used in equation (2.36) for the calculation of the standard deviation of the flight times (see figure 2.7).

The Reflectron The optimization of the reflectron is done in a similar way like the optimization of the accelerator described before. One difference is that we use the optimized values for potentials and lengths which we obtained by the optimization of the accelerator. Another difference is that we must add to the equations of motion the equations valid for the ion motion inside the reflectron. The reflectron acts as an ion mirror [112; 156]. The improved version of the one stage reflectron is the two stage reflectron. The first stage acts as a slow down region (deceleration stage), whereas the second stage acts as the soft reflection region (reflection stage, see figure 2.6). The deceleration stage of the reflectron can be treated as an "acceleration" stage with negative acceleration. However for the second stage we must consider the kinetic energy and the penetration depth of the incoming ion to obtain the turning point. In that case we make use of kinetic energy conservation related to the potential energy given by the turning

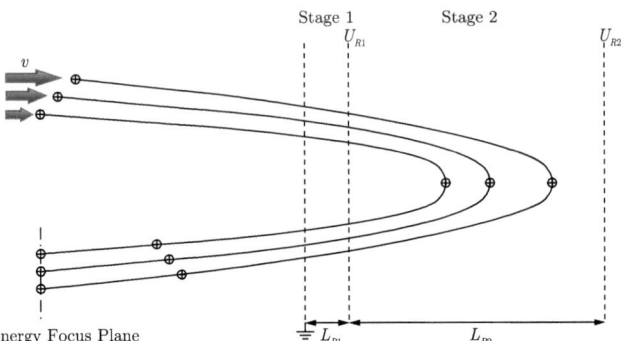

Figure 2.6 Shown is the schematic principle of energy focusing with a two stage reflectron. Incoming ions exhibit different energies (velocities) given by source conditions. The ions with higher kinetic energy penetrate deeper into the second stage than slower ions with the same mass to charge ratio. Thus the slower ions obtain a head start and can compensate the velocity difference. So at the energy focus plane where a detector would be placed ideally every ion will arrive at nearly the same time-of-flight.

point of the ion trajectory in the reflection stage. In the following we can write for the ion velocity v_{R1} after deceleration in the first reflectron stage with the length L_{R1} and the potential U_{R1}:

$$v_{R1}(x_s, v_0) = \sqrt{\frac{(L_1(U_0 - U_{R1}) + (U_1 - U_0)x_s)z}{L_1}} \pm v_0 \qquad (2.43)$$

The velocity v_{R1} after deceleration in the first reflectron stage only depends on the kinetic energy and not on the length of the stage L_{R1}. So it is obvious to write for the time the ion will need to pass the first stage of the reflectron:

$$t_{R1}(v_3, v_{R1}) = \frac{2L_{R1}}{v_3(x_s, v_0, v_1, v_2) + v_{R1}(x_s, v_0)} \qquad (2.44)$$

With the approach of energy conservation we can calculate the time, the ion will need to reach its turning point in the second reflectron stage (ion kinetic energy equal to the potential energy) [129; 135; 157; 158]:

$$t_{R2}(v_{R1}) = \frac{2L_{R2}\left(\frac{q((U_1 - U_0)x_s + (U_0 - U_{R1})L_1)}{L_1} \pm \frac{mv_0}{2}\right)}{qv_{R1}(U_{R2} - U_{R1})} \qquad (2.45)$$

where L_{R2} is the length of the second reflectron stage and U_{R1} the potential applied to the second reflectron stage. To obtain the total flight time of the ion one must add to equation (2.35) the flight times of the ion required for passing the first and second reflectron stage (2.44) and (2.45). However, one must keep in mind that these times must be multiplied by a factor of two to take into account that the ion passes the reflectron stages two times due to reflection. Additionally the distance from the reflectron entrance to the detector must be added to the field free drift length. With this addition equation (2.35) can be written in the form:

$$t_{\text{all}} = \sum_{i=1}^{d} t_i + \sum_{i=1}^{n} 2 t_{Ri} \qquad (2.46)$$

with n reflection stages or multiple reflection systems [159; 160] This new equation can then be defined in R as a new function, which can be optimized in the same way like the linear TOFMS described before.

2.2.6 The Optimization Procedure

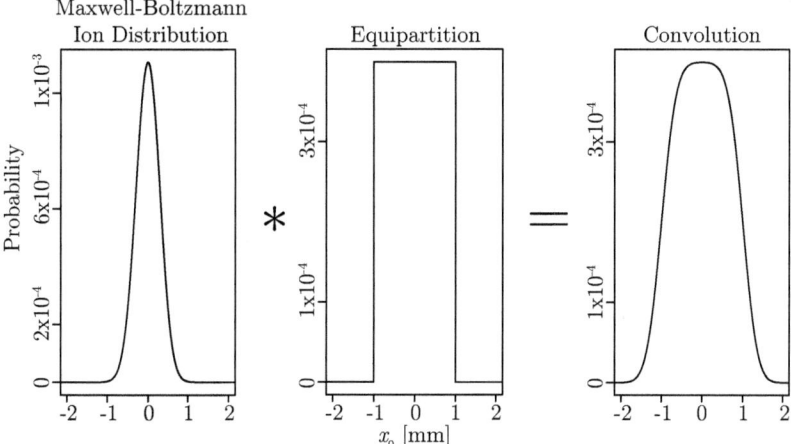

Figure 2.7 Convolution of the space dependent Maxwell-Boltzmann ($T_\perp = 1$ K) distribution with a step-function approximating the skimmer shape with a ⌀ = 2 mm. The resulting function can be used in equation (2.36) as the probability distribution $p(x_0, i)$ for ions starting at a position x_0.

The optimization programs are written in "R-language". All written programs have generally the same structure. Therefore we will describe at first the fundamental structure of the optimization process. At the beginning of the program the constants such as the estimated beam temperature $T_\perp$, elementary charge e, ion mass m, Boltzmann constant k_B and extraction potential U_0 are set to the known values. Secondly, the dimensions of the acceleration stages L_1, L_2, L_3 and L_D (see figure 2.5) are set to the desired values. To optimize a Wiley McLaren design one must just set L_3 to zero. To obtain their starting positions x_0, 10001 ions (for symmetry reasons) are arranged on a 4 mm line (estimated maximum beam width) centered around $L_1/2$. The velocity in extraction direction is then calculated by equation (2.38). In the next step the Maxwell-Boltzmann distribution of the ions is calculated with equation (2.41) for an estimated beam temperature of 1 K and convoluted with a step-function (square-shape estimation) approximating the skimmer inlet diameter (e. g. $\varnothing = 2$ mm). The resulting distribution probability is later used in equation (2.36) for the calculation of the standard deviation in flight times. An example for such a distribution calculation is depicted in figure 2.7. In the following all equations of motion (2.28) – (2.34) are collected in the overall flight time of the ion represented by equation (2.35). In that case the standard deviation given by equation (2.36) is used to define a new function in R, which is then optimized by the implemented function "optim". For the optimization process the starting values, the lower boundary, the upper boundary and some controlling parameters like maximum iterations are entered. During the optimization the function "optim" searches in the parameter space defined by the lower and upper boundary for the parameter set which delivers the minimum value (lowest flight time errors). This minimal value is then used in equation (2.37) for calculation of the optimized resolution. The function "optim" works with multidimensional parameter sets. For the optimization of the drift length L_D for example, this length must be included in the function which will be optimized as a parameter. So in that way second or higher order focusing is possible. However, one has to "play" with the option parameters of the function "optim" to verify that a global and not a local minimum in the parameter space is located by the optimization process.

2.3 Metastable Decay

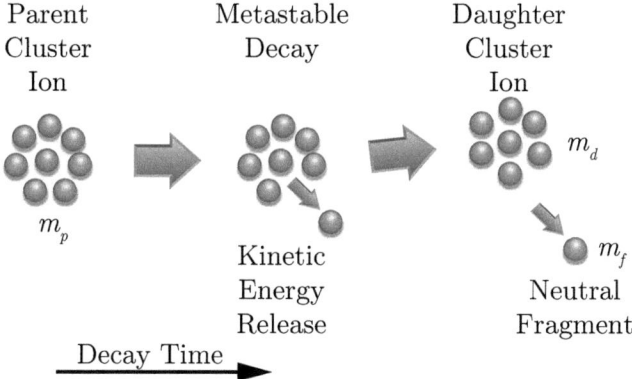

Figure 2.8 Shown is a schematic diagram of metastable decay. After a decay time (τ) the excited metastable parent cluster ion with mass m_p spontaneously decays via unimolecular dissociation to a daughter cluster ion with mass m_d and a neutral monomer with mass m_f.

Metastability is known in physics as a state of apparent stability that is capable of changing to a more stable state when subjected to perturbation[4]. Supersonic jet expansion yields molecular beams of vibrationally and rotationally "cooled" weakly bound neutral clusters [37]. However, as mentioned before a prior ionization of these neutral clusters is required for mass spectrometric investigations (see also subsection 3.2.2). In that case e. g. an energetic electron beam (electron ionization, EI) or an energetic light source (Laser ionization) is used for the ionization of the neutral clusters. According to these ionization methods an electron or a photon interacts with the initially cold neutral cluster (these processes are very fast compared to the following processes). Due to the interaction of the electron or photon with the neutral cluster excess energy above the ionization threshold is imparted to the cluster ion [161]. Energy exchange of the newly generated excited cluster ion with the environment is "not possible" due to the molecular beam conditions (depending on the distance between the expansion valve and the ionization source, see subsection 4.2.5). Therefore the excited cluster ion cannot establish a thermal equilibrium (at greater distance between the expansion valve and the ionization source). Hence the ionization imparted excess energy is redistributed into different vibrational modes of the "hot" cluster. Within this

---

[4]"metastable", The Oxford English Dictionary, 2nd ed. 1989, OED Online, Oxford University Press, 2000, http://dictionary.oed.com/cgi/entry/00307537

metastable state the excited cluster can undergo evaporative metastable dissociation or not (see figure 2.8). Depending on the excitation energies Beu et al. [76] mention three different fragmentation regimes. Mainly large and small fragments are formed at low excitation energies. In that "evaporation" regime the excited "hot" metastable cluster evaporates a small number of single particles. At very high excitation energies the parent cluster fragments completely into small pieces which is known as the "shattering" regime (comparable with the shattering of clusters upon surface impact, see section 2.4). The third regime is an intermediate regime between these two other regimes. In EI mass spectrometry almost exclusively the unimolecular dissociation is observed, e. g. in the form:

$$(A)_n \xrightarrow{h\nu, e^-} (A)_n^+ \longrightarrow (A)_{n-1}^+ + A \qquad (2.47)$$

Regarding the unimolecular dissociation, metastable decay can occur spontaneously even on a μs-time scale. In the case of clusters the properties change noticeably as a result of the decomposition. These changes include the sublimation energy, decay rate and internal temperature [162]. First occurrence of metastable decay was reported by Stace and Shukla [163] for the metastable dissociation of carbon dioxide cations (see 2.47). However they find that there is no evidence that the cluster ions lose more than one monomer unit within the decay. These measurements were carried out with a single focusing sector type device. Many studies of other molecules and clusters followed this prior work. Comprehensive overviews of the field of metastable decay were given in various reviews (e. g. [15; 161; 164; 165]). The binding energy of a molecule can be determined by statistical models e. g. the modified quasi equilibrium theory (QET)/Rice-Ramsperger-Kassel (RRK) statistical model introduced by Engelking [166]. According to Engelking the binding energy of a molecule in the cluster can be calculated by the measurement of the average kinetic energy release (KER) and the metastable lifetime (dissociation rate). Here the basic idea is that metastable clusters are observed only if their lifetimes fall within the experimental observation time-"window". In that case in mass spectra besides the "parent" cluster ion peak $(A)_n^+$ a "daughter" cluster ion peak $(A)_{n-1}^+$ appears. The daughter cluster ion peak is the result of the metastable decay reaction given in (2.47). In the case that no kinetic energy would be released during the metastable decay reaction the daughter cluster ion peak would have the same width as the parent cluster ion peak. However, taking any KER in the reaction into account will change the peak shape of the daughter cluster ion peak. Assuming Gaussian peak shapes the KER can be extracted from the FWHM of the daughter cluster ion peak (related to the width of the parent cluster ion peak) [71; 167; 168]. Most of the studies of dissociation dynamics of metastable cluster ions have been carried out with double focusing two-sector field mass spectrometer devices (e. g. as in [168; 169]). Later Castleman and coworkers introduced a new technique to derive KER and decay fractions with a reflectron TOFMS (reviewed in [167]). They employed the reflectron to separate daughter and parent ions in order to measure KER and decay fractions of dissociating cluster ions in the field free region of the TOFMS (see also subsection 4.3.1, figure 4.49). Principally the technique is based on the kinetic energy difference between the parent cluster ion and the metastable decay product, the daughter cluster ion. According to the kinetic energy difference these ions possess different penetration depth in the reflectron and thus TOF's. As mentioned before KER can be determined by peak shape analysis. The mass of the daughter cluster ion can be deduced by kinetic energy analysis (see equation 4.4 in subsection 4.3.1). Alternatively Stairs et al. [170], L'Hermite et al. [171] and Gilmore et al. [172] described similar methods for the determination of the daughter cluster ion masses and possible metastable decay channels with a TOFMS reflectron.

2.4 Cluster-Surface Interactions

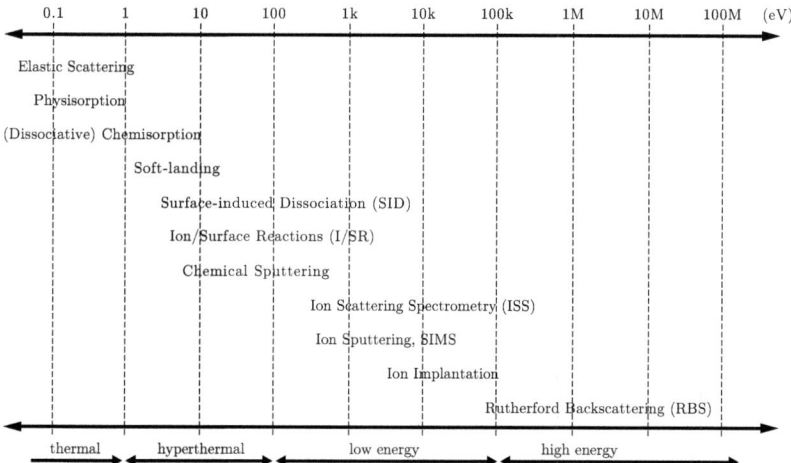

Figure 2.9 Fundamental particle-surface interaction processes observed from thermal to high collision energies. Indicated are the energy regions where these processes are typically observed [173].

Clusters bridge the gap between atoms (or molecules) and the condensed bulk matter phase. The evolution of microscopic to macroscopic properties generally shows no linearity for small cluster sizes [4–8]. Hence a size dependent study of different cluster systems is demanded. One possible analysis method is the investigation of the interaction of clusters with solid surfaces under well known conditions (e.g. cluster size, interaction energy, surface structure and so forth). Therefore up to today the interaction of different cluster systems with well defined solid surfaces has attracted much interest. Comprehensive reviews of this complex and wide field exist [16; 17; 173–176]. A first rough restriction to domains of cluster types can be made by the classification of the interacting cluster in either neutral [177–190] or charged ionic species [174; 191–224]. Besides these experimental-based efforts many theoretical calculations were performed to understand or predict some experimental observations [225–243].

Depending on the collision energy (and energy per atom or molecule), cluster size and cluster species different processes can be observed. In literature scattering experiments are roughly and arbitrarily divided into four different regimes defined by the selected collision energy [173; 244]. The lowest energy regime termed as the thermal range involves ions with kinetic energies below 1 eV. Generally molecular beams produce particles in this energy range. Additionally by seeding the sample gas with different ratios and different carrier gases the velocity and thus the collision energy of the particles in the beam can be precisely adjusted [62; 64; 245–249]. In this case collision energies from milli-electron-volt (meV) up to several eV can be achieved depending on the cluster size of the molecular clusters involved. The hyperthermal range covers the energy range above the thermal 1 eV up to 100 eV. The low energy range and the high energy range cover the range between 0.1–10 keV and up to Mega-electron-volt (MeV) respectively. The mentioned energy regimes and the processes observed within these regimes are depicted in figure 2.9. The chemically most interesting energy regime is the hyperthermal energy regime with $\sim$ 1 eV to about 100 eV. Within this regime the collision energy is comparable or greater than typical chemical bond and cluster-binding energies of the colliding particles (see table 2.1). Besides, the ion's translational energy is large enough to cause bond cleavages or fragmentation of the projectiles; however, it is not so large as to completely transform and so obscure the chemical nature of the projectile-surface collision pair. Additionally at hyperthermal energies new chemical bonds can be formed as well as broken due to impact induced intra cluster reactions.

Here we will give a brief overview of the processes related to cluster surface collisions at hyperthermal energies (after ref. [17]). Upon the surface impact the translational energy is partly transfered to internal energy of the cluster. Compared to atom-surface collisions the molecule-surface and cluster-surface collisions provide the vibrational excitation as a new channel where the impact energy can be transfered. Cleveland et al. calculated with molecular-dynamics simulations for the surface collision of an argon cluster consisting of 4000 atoms that the effective "temperature" and "pressure" reach 4000 K and 10 GPa, respectively (1.9 eV per argon atom, on NaCl(001) surface) [226]. Such extremely high compression and energy densities cannot be achieved by atomic ion impact. Regarding these extreme conditions, cluster-surface interactions provide an opportunity to observe novel and unique processes.

Surface Deposition On the other hand it is possible to deposit size selected clusters intact on a surface at sufficiently low collision energies [250; 251]. This process can be regarded as a new method to prepare nanostructured surfaces.

Non-Dissociative Scattering At low collision energies (compared to the binding energy of the cluster) or quasi elastically scattering from the surface the colliding cluster can survive the collision without dissociation. In that case the binding energy of the cluster is higher than the sum of the collision energy and the interaction energy between the cluster and the surface. Hence the excess energy gained by the collision is accommodated by the vibrational degrees of freedom available in the cluster. The number of vibrational degrees of freedom increases with the size of the cluster. Therefore large clusters such as e. g. the fullerene C_{60}^- can survive collisions on a Si(100) surface with less than 170 eV (2.8 eV per carbon atom) collision energy [252].

Dissociative Scattering Dissociative scattering occurs in the cases where the collision energy exceeds the binding energy of the cluster. In that case the excess energy gained by the collision process is too high to be stored in the cluster as internal energy (see also section 2.3 about metastable decay). This leads to the dissociation of the cluster which can take place in various ways:

Impact Dissociation by Evaporation In that case a "big" however slightly bounded cluster (e. g. Lenard-Jones or van der Waals) interacts with the surface. Due to the different collision induced processes the cluster can be schematically divided into three zones [253]. Atoms near the surface interact most strongly with the surface. Therefore these atoms remain on the surface as atomic adsorbates. Atoms on the top of this zone are divided into two different zones. One zone located in the center of the cluster and one zone located around the center up to the boundary. In the center the atoms can evaporate directly without any flow velocity. Atoms on the other hand in the zone of the cluster borders acquire a lateral flow velocity, glide on the argon adsorbates and evaporate directly giving a broad fragment angular distribution (see also [234; 243]). The resulting process is comparable with the well known Leidenfrost phenomenon.

Shattering As mentioned before the excess energy gained by the surface collision can exceed the binding energy of the cluster. In that case the cluster evaporates fragments to "cool" down. However, it was reported that the cluster *shatters* to small pieces typically monomers when the internal energy of the cluster exceeds a critical value [197; 203; 208; 209; 228]. This process resembles a phase transition which occurs after reaching a certain excitation (collision) energy. Simulations and experiments [208; 228; 229; 237; 241] showed that the shattering event is much faster (< 1 ps) than the evaporation process. In contrast to the fast shattering event cluster surface collision induced evaporation occurs delayed after a longer time period (typically ≈ 100 ps).

Fission and Evaporation One of the main factors which influence the outcome of cluster surface collision experiments is the nature of the interatomic interaction of a cluster. Silicon cluster ions which collide with a silicon surface tend to split into almost equally pieces which is known as fission [193]. Compared to this process observed for silicon cluster ions antimony and bismuth cluster ions show unimolecular dissociation of stable neutral clusters (depending on the parent cluster size) [220].

Cleavage Beck et al. observed for the impact of $Na_n F_{n-1}$ cluster that the cluster-surface scattering is highly inelastic (up to 35% of the incident kinetic energy is dissipated in the internal heating of the cluster). At low collision energies these cluster ions show impact induced cleavage with a crossover to impact induced evaporative decay observed for impact energies higher than 1 eV per atom [191].

Intracluster Reactions

Mechanical Bond Splitting It was observed that a diatomic molecule ion I_2^- embedded in a CO_2 cluster can be split mechanically by surface collision on a silicon surface [204]. In that case one of the CO_2 molecules in the cluster act as a molecular "wedge" during the surface collision of the cluster.

Shock Wave Induced Dissociation Embedded reactant molecules can be highly and impulsively excited by the impact of a large cluster containing the reactant molecules. Sheck et al. [229] showed with molecular-dynamics simulations that such high and impulsive energy transmission can occur in impact induced shock waves (on a nanometer scale).

Intracluster Four-Center Reactions In chemistry it is improbable that a four-center reaction proceeds under ordinary reaction conditions. Such reactions have high energy barriers and are generally accompanied by large energy releases. However, Raz and Levine [228] have predicted in a theoretical work a cluster-surface collision induced four-center reaction between N_2 and O_2 embedded in a large rare-gas cluster. Experimentally Christen et al. [210; 213] showed that four-center chemical reactions can be induced between alkyl halide molecules by cluster-surface collision on a p-type diamond covered silicon surface. The experiments showed that the reaction probability increases with the cluster size of the colliding cluster as predicted by the theoretical calculation of Raz and Levine.

Electronic Interaction

Electron Transfer to the Surface Cluster ion surface collisions involve to some extent charge transfer between the cluster ion and the surface. Regarding a cluster anion, the electron of the cluster can be transfered during the surface collision to the surface. Such processes depend on the electronic structure of the cluster anion and the electronic structure of the surface. For e. g. the $I^-(CO_2)_n$ cluster anions the electron transfer depends critically on the CO_2 solvation structure of the cluster anion and decreases with increasing cluster size [212].

Electron Emission Besides the impact induced unimolecular decay of the cluster an impact "heated" cluster can emit electrons. In that case the energy gained by the cluster-surface collision is higher than the binding energy of the electron in the cluster. Many factors influence the rate of electron emission e. g. collision energy, cluster size (degrees of freedom), the electronic and geometrical structure of the cluster.

Secondary Emission from the Surface Secondary electron or ion emission from a surface induced by the impact of an energetic particle is well known for primerion beam impact in the keV collision energy regime [254]. For cluster ions similar processes become dominant when the collision energy reaches comparable high values (≈ 1 keV) [199]. The interest in these processes increased in time due to the potential of using cluster ion beams in secondary ion mass spectrometry (SIMS) [255].

Chapter 3
Experimental Setup

> *"Tell me... And I Forget,*
> *Teach me... And I Learn,*
> *Involve Me... And I Remember."*
> BENJAMIN FRANKLIN (1706–1790)

3.1 Assembly and Vacuum System

The assembly of the experimental setup used for this work is depicted in figure 3.1. This apparatus was designed and build to investigate the interaction of size selected molecular clusters with solid surfaces at hyperthermal energies [256]. The assembly consists mainly of three differentially pumped ultra-high vacuum (UHV) chambers (Pink GmbH, Germany). Completely hydrocarbon-free pumping stages are in use to avoid the disturbance of pumping oil on the investigation of the cluster-surface interactions. Therefore every chamber is evacuated by pumping stages consisting of corrosion-resistant, magnetically levitated turbomolecular drag pumps (Pfeiffer Vacuum GmbH, TMU 1000MPCT, Germany) backed by chemically persistent diaphragm pumps (Vacuubrand, GmbH, MD 4BRL, Germany). The first chamber (diameter $\varnothing \approx 420$ mm) serves as the expansion and ionization chamber. It contains a pulsed nozzle for the formation of the supersonic cluster beam. A nozzle mounted or a flange mounted electron gun can be used with variable electron energy for the ionization of the generated clusters (see subsection 3.2.2). The core of the molecular beam is extracted by passing through two conical homebuilt skimmers ($\varnothing = 3$ mm and $\varnothing = 2$ mm, 4 and 6 in figure 3.1). The second chamber is just used for the differential pumping and reduction of the gas load produced in the expansion chamber. Thus in the third chamber UHV conditions can be remained even at high gas loads. Base pressure in all chambers is well below 5×10^{-7} Pa without baking. During operation in the third chamber this pressure increases slightly, but not exceeding 1×10^{-6} Pa over the day during the measurements. The third chamber contains the home built Re-TOFMS with a mass gate and reflectron collider. With this device size selection prior to impact, deceleration to the desired collision energy and subsequently mass and energy analysis of the collision products can be per-

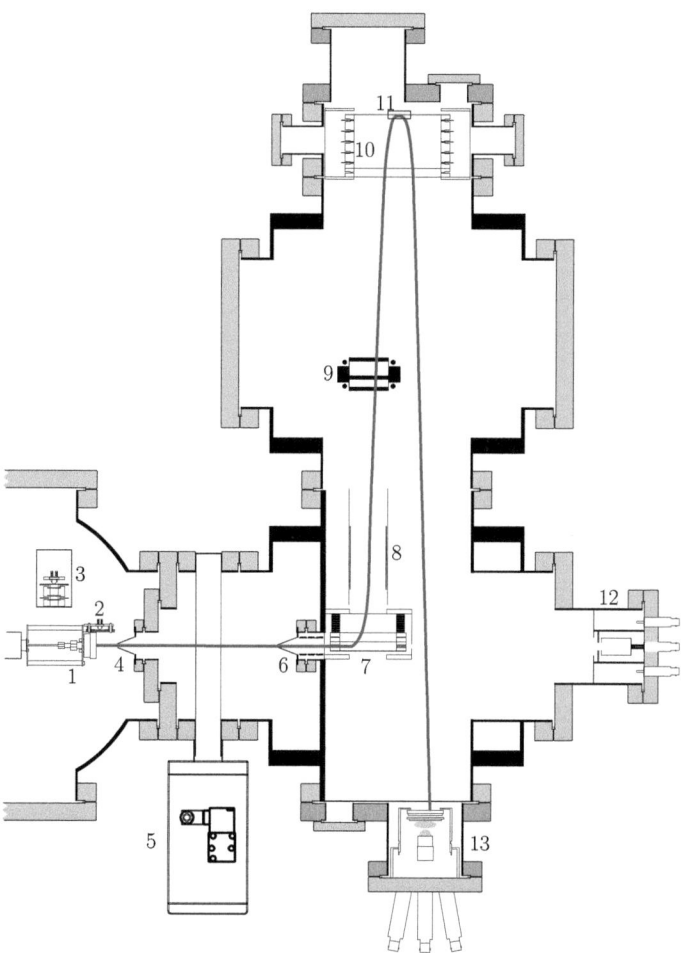

Figure 3.1 Drawing of the experimental setup for cluster-surface interaction studies after [256]: (1) Pulsed nozzle mounted on a xyz-translator stage (not shown). (2) Nozzle mounted electron gun. (3) Flange mounted electron gun. (4) Conical skimmer (can be biased). (5) Gate valve. (6) Conical skimmer. (7) Three-stage TOF-accelerator. (8) TOF-deflector. (9) Pulsed interleaved comb mass gate. (10) Two-stage reflectron. (11) Heatable surface mounted on the back of the reflectron. (12) Faraday-cup and retarding field analyzer. (13) MCP-Detector with rotatable retarding field analyzer (not shown).

formed. Therefore in front of the detector (13 in figure 3.1) a rotatable retarding field energy analyzer was implemented which can be swung in and out from the beamline (for retarding field principle see 3.3.1). A detailed description of the design considerations, development, optimization and simulation of the Re-TOFMS parts will be given in the following chapter 4 (for pictures of the Re-TOFMS components see appendix A).

3.2 Cluster-Ion Generation

3.2.1 Pulsed Nozzle

Clusters were formed by the pulsed supersonic molecular beam expansion with a solenoid nozzle. The jet source is a high pressure and temperature valve which is additionally suitable for cryogenic operation. The design of the jet source is based on the valve design introduced by Even et al. [30]. The valve is designed and customized for the experimental needs by Nachum Lavie (School of Chemistry, Tel-Aviv University, Israel). The source can be operated with high stagnation pressures $p_0 = 0.2$–12 MPa and with a maximum repetition rate of 25 Hz. A syringe pump (Teledyne Isco, Inc., USA) is used to control the stagnation pressure. Stagnation temperature ($T_0 = 225$–425 K) is controlled by a highly dynamic temperature system (Julabo Labortechnik GmbH, LH85, Germany). A custom built active temperature controlling experimental setup is used for the stabilization and control of the stagnation conditions ($\Delta T_0 < 30$ mK and $\Delta p_0 < 2.9$ kPa) [257]. The novel miniaturized valve in use has the following advantages to past designs. The high stagnation pressure makes it possible to cool large aromatic molecules to less than 1 K. Further the gas flow to the vacuum chamber is minimized due to the short opening time of the pulsed valve ($\sim$20 μs). Additional to this, by the short gas pulse a packet of fast moving clusters is formed which is well utilized by TOFMS devices.

3.2.2 Electron Guns

Supersonic nozzle expansions generate clusters with a broad size distribution. Prior ionization is required for the detection of these clusters with a mass spectrometer. In the field of mass spectrometry the electron-impact ionization (EI) is an established universal ionization tool. Contrary to photoionization with lasers, EI is a versatile ionization tool which is also amenable to miniaturization [258]. Photoionization sources are more selective than EI due to the more or less fixed wavelength of lasers. Most of the molecules have their strong one photon absorption for ionization in the ultraviolet regime [131]. However this can lead to heavy fragmentation of clusters which is in general not desired in most cluster experiments. Therefore low energy EI is better suited to get improved control over the ionization induced dissociation and fragmentation effects. Besides these characteristics EI also allows the generation of negative and positive species depending on the electron energy and the nature of the neutral sample [11; 259]. For most molecules and the rare gases the maximum of the ionization probability is located between 50–100 eV [260]. The ion yield generated by EI is proportional to the available electron current and energy. At the same moment the electron current steeply increases with increasing electron energy [261; 262]. The highest possible electron current is limited only by space charge in the beam itself [263]. Due to the dependence of the degree of clustering from the nozzle to electron-gun (e-gun) distance [264] two EI sources were used in the present work (see figure 3.1). One of the guns has a slim design and is mounted on the nozzle. This gun is used for the formation of big sized clusters by ionized germs. With this e-gun ionization occurs directly at the nozzle exit. In that case the ions are generated during the cluster generation process in which ionized cores grow further by successive attachment of monomers [264]. The other positive aspect of such a configuration is that fragmentation by the ionization process is reduced. Due to ionization during the nucleation of the clusters, "hot" clusters have time to be cooled down by collisions in the beam. On the other hand ionization at larger distance to the nozzle produces smaller cluster-ions. By the variation of the electron source to nozzle distance the size distribution of the cluster-ions shift. Therefore a second flange mounted e-gun can be used for ionization at different nozzle to e-gun distances. The background to operate two different ion sources with different distances to the nozzle is to maximize the yield of a certain desired cluster size. The designs and configurations of the two applied electron guns (2 and 3 in figure 3.1) are depicted in the appendix B

3.3 Detection Sytems

3.3.1 Faraday Cup

Faraday cups are well known devices for the detection of charged particles [265]. Due to their sturdy and facile construction these devices are widely in use. A basic Faraday cup consists of two cylinders. An outer cylinder contains the second inner nested cylinder (the cup). The outer cylinder is generally held at ground potential and is insulated from the inner cylinder. From an aperture on the outer cylinder the charged particles (ions or electrons) enter into the inner cup which is referred to as collector. The measurement of ion fluxes to the collector generally requires amplification and precautions for noise reduction. Secondary electron emission due to ion impact [254] is a general problem for Faraday cups. Therefore in most cases the surface of the collector is rough or structured. A rough or structured collector increases the probability to reabsorb the emitted secondary electrons avoiding them to leave the cup and cause an error in the measurement.

This can be enhanced by using a negatively biased mesh in front of the entering aperture. In cluster science the capability of measuring cluster size distributions with the Faraday cup is of interest. For this request the basic Faraday cup used in the current work (see figure 3.1) was expanded by an energy analyzer [266; 267]. The energy analysis is done with a retarding field technique (see subsection 2.1.3). For this purpose two grids are used. The first one is kept at ground potential and the second one is kept at negative or positive potential depending on the measured particle charge. An incoming particle must overcome this potential barrier to get into the cup. Hence only the particles can be detected which possess a kinetic energy that is higher than the retarding field potential barrier. With the measurement of the beam current in dependence of the retarding field the size distribution of the detected particles can be derived. Additionally for supersonic beams the velocity of the expanded gas can be calculated by the required time-of-flight from the nozzle to the cup. The current from the cup is amplified by a current amplifier LCA-4K-1G (Femto, Germany, 4 kHz bandwidth, 1 GV/A amplification). With its high sensitivity the amplifier allows time resolved measurement of the ion or electron current by a digital storage oscilloscope (LeCroy, Switzerland).

3.3.2 Microchannel Plate Detector

Time-of-flight mass spectrometer devices require fast and sensitive detectors. In general the response time of the employed detector defines the upper limit of the maximum possible resolution. Ion detection with microchannel plate (MCP) detectors [268] evolved to the default detection method in TOFMS beside many other detection methods. MCP detectors offer fast output signal rise times (below 500 ps) and large detection areas (up to $\varnothing = 40$ mm). Signal gain is improved by the use of two MCP detectors one behind the other also known as the Chevron configuration [269]. The detector used in this work (see figure 3.1) is a high-speed bipolar MCP hybrid device with a sensitive are of $\varnothing = 25$ mm (Burle, Inc., USA). It consists of a MCP for ion-to-electron conversion and amplification, a scintillator electron-to-photon conversion surface and a photomultiplier tube (PMT) detector. The MCP detector offers the post-acceleration of both positive and negative ions with up to 10 kV which is important for the detection of large and heavy species [133; 270].

3.4 Electronics

The main task of TOFMS electronics is to measure the time-of-flight of charged particles. Therefore a precise time-base with high accuracy and resolution is needed. The time-base of the present setup is controlled by a DG645 digital pulse-delay generator (Stanford Research Systems, Inc., USA). The DG645 has a resolution of 5 ps and an accuracy of 1 ns. The nozzle opening is triggered by a 20 μs long TTL pulse from the DG645. Optional to the continuous e-gun operation, delayed to the nozzle opening one of the electron guns can be pulsed. The 10 MHz output of the DG645 is used to time synchronize two additional pulse-delay generators DG535 (Stanford Research Systems, Inc., USA). The two DG535 devices generate the delay for ion extraction (accelerator), mass selection (mass gate) and data acquisition (digital storage oscilloscope). Alternatively, for time-resolved ion counting the DG645 delivers the start and stop signal to a 4 GHz multiple-event time digitizer P7887 (FAST ComTec, GmbH, Germany) with 250 ps time resolution. Pulsed operation of the TOFMS components is controlled by up to six fast high voltage transistor switches capable to switch up to 8 kV (HTS 81-06-GSM, Behlke Electronics, GmbH, Germany). High voltage up to 6 kV is provided to the push-pull switches by virtually ripple-free (< 5 mV) high voltage power supplies NHQ (iseg Spezialelektronik, GmbH, Germany) which can be optionally computer controlled set within ± 40 mV.

Chapter 4
Results and Discussion

> *"We have a habit in writing articles published in scientific journals to make the work as finished as possible, to cover up all the tracks, to not worry about the blind alleys or describe how you had the wrong idea first, and so on. So there isn't any place to publish, in a dignified manner, what you actually did in order to get to do the work."*
> RICHARD PHILLIPS FEYNMAN (1918–1988)

4.1 Design, Numerical Optimization, Simulation of the Reflectron-TOFMS

4.1.1 Numerical Optimization

The TOFMS-Accelerator In this subsection the results obtained by numerical optimization (see 2.2.5) of the TOFMS accelerator will be summarized. To get started with the numerical optimization process at first a generic Wiley-McLaren type two stage accelerator was calculated. To avoid complexity and gain an overview of the optimization process an equipartition distribution of starting ions with no velocity distribution was assumed. A representative result of such an optimization calculation is depicted in figure (4.1). The flight length L_d was increased successively beginning from 0 mm up to 600 mm. In this case for every point in figure (4.1) the optimization process yields the optimized value of U_1 and the resulting optimum resolution calculated for fixed values of U_0, L_1, L_2 and L_d (*first order space focusing*). It can be observed that the optimum resolution increases with increasing L_d reaching a maximum value at around $L_d = 114$ mm and falls with further increase of L_d. According to this local resolution maximum at $L_d = 114$ mm second order space focusing will be achieved by setting $L_d = 114$ mm. This result was later approved by adding L_d as an additional parameter besides U_1 in the optimization process (*second order space focusing*). In addition to that these results were also approved by the analytical optimization calculation according to Even and Dick [146] which resulted in nearly the same values for L_d and U_1 ($L_d = 114.13$ mm and $U_1 = 3142.06$ V discrepancies

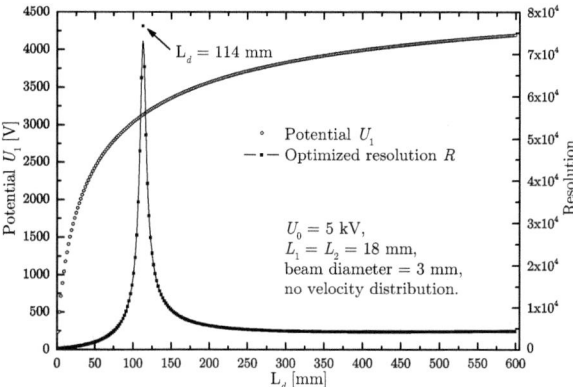

Figure 4.1 First order space focusing with a Wiley-McLaren type two stage accelerator calculated by numerical optimization. Both stages are 18 mm long, repeller extraction voltage fixed at 5 kV. It was assumed that the ions start with no velocity distribution along the beam diameter of 3 mm. The optimum value of U_1 (o) and the resulting resolution (-•-) for increasing drift length L_d.

below 1%). To take in to account the supersonic beam source properties the optimization process was further improved by the addition of an ion starting velocity distribution and a starting position probability distribution (see 2.2.5). It must be noted here that in that case to find an analytical solution is barely possible. Additional numerical optimizations showed that the lengths of the acceleration stages influence the position of the space focus plane dramatically. In this sense the question arises which configuration of the acceleration stage lengths fulfills the geometrical requirements given by the experimental setup. The flight distance from the molecular beam inlet up to the detector (linear configuration) is nearly 600 mm long plus 100 mm reserved space for the reflectron. A mass gate will be placed at half distance (around $\approx$ 300 mm) to the detector which should not disturb the reflected ions when later a reflectron is available. This would limit the accelerator length plus the flight distance to the mass gate to roughly 300 mm. Due to the fact that the resolution profits more from the length L_d this length should be long as possible and the whole accelerator length $L_{\text{accel}} = L_1 + L_2$ short as possible. Therefore we decided to limit the accelerator length L_{accel} to roughly $L_{\text{accel}} = L_1 + L_2 \approx 50$ mm and the field free flight path length L_d to $L_d \approx 250$ mm. The first question that arises was if it is possible to fulfill this requirement with the standard two stage Wiley-McLaren configuration ($L_{\text{accel}} = L_1 + L_2 \approx 50$ mm with an optimized $L_d \approx 250$ mm). Therefore a new approach was made. We calculated for the whole accelerator length L_{accel} between 30 mm up to 90 mm the resulting optimal flight length L_d to the space focus plane. Hence, the length

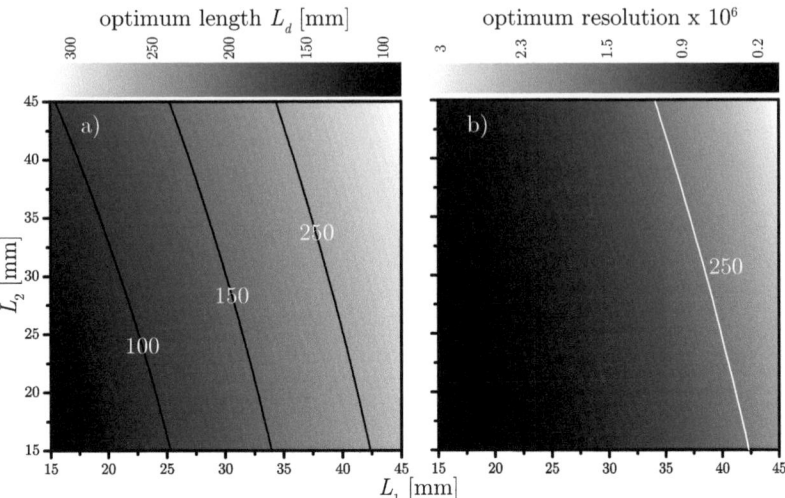

Figure 4.2 Second order space focusing with a Wiley-McLaren type two stage accelerator calculated by numerical optimization. The lengths of both acceleration stages are varied between 15 mm up to 45 mm. Repeller extraction voltage is fixed at 6 kV. It is assumed that the ions start with a Maxwellian velocity distribution according a beam temperature of $T_\perp = 1$ K along the beam diameter of 3 mm. **a)** The optimum position of the space focus plane L_d in dependence of the acceleration stage length L_1 and L_2. **b)** Optimum resolution obtained for the values depicted in a).

of the first acceleration stage L_1 and the second acceleration stage L_2 was varied successively and the optimum flight length to the space focus plane was calculated by numerical optimization. The obtained results are depicted in figure (4.2) **a)** and (4.2) **b)**. For the Wiley-McLaren configuration the optimum $L_d = 250$ mm is obtained for larger L_1 and L_2 exceeding the overall geometrically desired limit of $L_1 + L_2 \approx 50$ mm. Highest resolution for $L_d = 250$ mm was obtained for an overall accelerator length of about 76 mm ($L_1 = 36$ mm, $L_2 = 40$ mm and $L_d = 250$ mm). Another trend which can be seen in figure (4.2) **a)** is that the resolution increases with shorter L_1 and increasing L_2. However, such values will exceed the geometrically determined accelerator length limit. Additionally, it is difficult to construct such long acceleration stages with perfectly homogeneous electric fields. Here the question arises if an accelerator with three stages will better suit the geometrical design considerations ($L_d \approx 250$ mm and $L_{\text{accel}} = L_1 + L_2 + L_3 \approx 50$ mm) than the Wiley-McLaren configuration. Therefore the influences of each acceleration

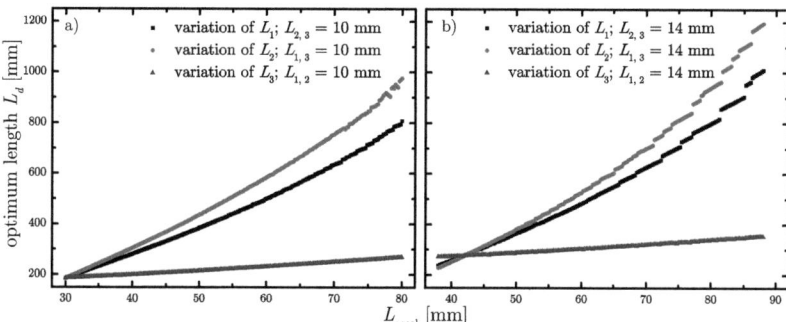

Figure 4.3 Alternate variation of the accelerator length L_i and its influence on the space focus plane distance L_d. It is assumed that the ions start with a Maxwellian velocity distribution according a beam temperature of $T_\perp = 1$ K along the beam diameter of 3 mm at 6 kV acceleration. **a)** Two of the acceleration lengths are fixed at the same value of 10 mm and the third length value is increased successively. **b)** Two of the acceleration length are fixed at the same length value of 14 mm the third length value is increased successively.

stage length on the space focus plane distance L_d were analyzed. In that sense alternately two of the accelerator lengths were hold at fixed values and the third one was increased successively. Two representative results of these calculations are depicted in figure (4.3). The optimization results in figure (4.3) show that each increasing accelerator length L_i increases the distance to the space focus plane L_d too. However, here the accelerator length L_2 has the greatest influence on the field free flight path length L_d. The influence of L_1 on L_d is quit lower than in the case of L_2. The length L_3 has the lowest influence on the optimum length of L_d which increases nearly linearly in comparison to the other accelerator stage lengths L_1 and L_2. The other value of interest is the optimized resolution for the optimum value of L_d. Analogously to the space focus distance the influence on the optimized resolution is depicted in figure (4.4). Since the resolution scales with the distance of the space focus plane the resolution R in figure (4.4) is scaled to L_d. The influence of each stage length to the scaled resolution is similar to the results depicted in figure (4.3). Again here the length L_3 has the lowest influence on the scaled resolution which increases again linearly with increasing L_3. However, the influences of L_1 and L_2 on the scaled resolution (R/L_d) behave quite differently than in figure (4.3). With increasing L_1 the scaled resolution increases as well reaching a saturation for very large values of L_1. Increasing the length L_1 dramatically enhances the scaled resolution in contrast to increasing the lengths L_2 and L_3. In the case of L_2 the increase of L_2 results at first in an increased scaled resolution which saturates faster than in the case of L_1. Further

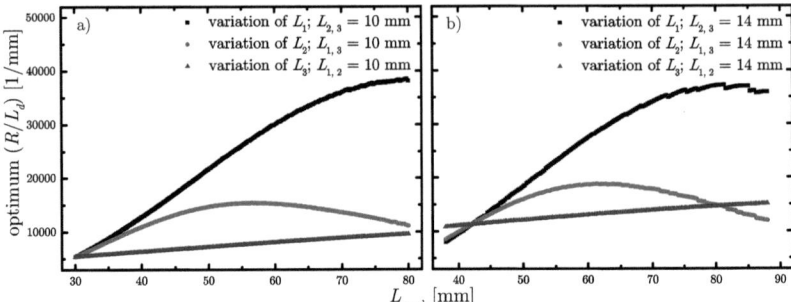

Figure 4.4 Alternate variation of the accelerator length L_i and its influence on the scaled resolution (R/L_D). It is assumed that the ions start with a Maxwellian velocity distribution according to a beam temperature of $T_\perp = 1$ K along the beam diameter of 3 mm at 6 kV acceleration. **a)** Two of the acceleration lengths are fixed at 10 mm the third length is increased successively. **b)** Two of the acceleration lengths are fixed at 14 mm the third length is increased successively.

increase in L_2 decreases the scaled resolution. This behavior can be explained by the larger influence of the acceleration stage length L_2 on the field free flight path length L_d. These results indicate a stronger influence of the two acceleration stage lengths L_1 and L_2 in contrast to the third acceleration stage length L_3. Hence additional calculations can be done focusing on the two acceleration stage lengths L_1 and L_2 which have a stronger influence on the optimum resolution R and the distance to the space focus plane L_d. In that sense we set the total length of the accelerator to the desired value $L_{accel} = 50$ mm and search the parameter set which delivers the highest resolution for $L_d = 250$ mm by the variation of L_1 and L_2. Due to the weaker influence of L_3 we define L_3 as $L_3 = 50$ mm $-(L_1 + L_2)$. The resulting contour plot of the optimum space focus distance L_d and resolution R is depicted in figure (4.5). The parameter set with an optimum space focus plane distance of $L_d = 250$ mm is displayed in figure (4.5) a) as a black line and in b) as a white line. Both lines for $L_d = 250$ mm show a linear dependence of the ratio between the length L_1, L_2 and thus L_3. Considering the resolution R in figure (4.5) b) it is again apparent that the first accelerator stage lengths L_1 has the largest impact on resolution. For $L_d = 250$ mm the best resolution (theoretical $R = 2.56 \times 10^6$) is available for the parameter set with $L_1 = 14.6$ mm, $L_2 = 10$ mm and $L_3 = 25.4$ mm. Contrary to this result the lowest possible optimized resolution (theoretical $R = 2.25 \times 10^6$ extracted from the contour plot) for $L_d = 250$ mm is given for the parameter set with $L_1 = 10$ mm, $L_2 = 13.7$ mm and $L_3 = 26.3$ mm (50 mm - 23.7 mm). This resolution value is nearly 6% lower than the calculated highest value for $L_d = 250$ mm for the three stage accelerator. However it must be kept in mind that these parameter sets are limited to the pa-

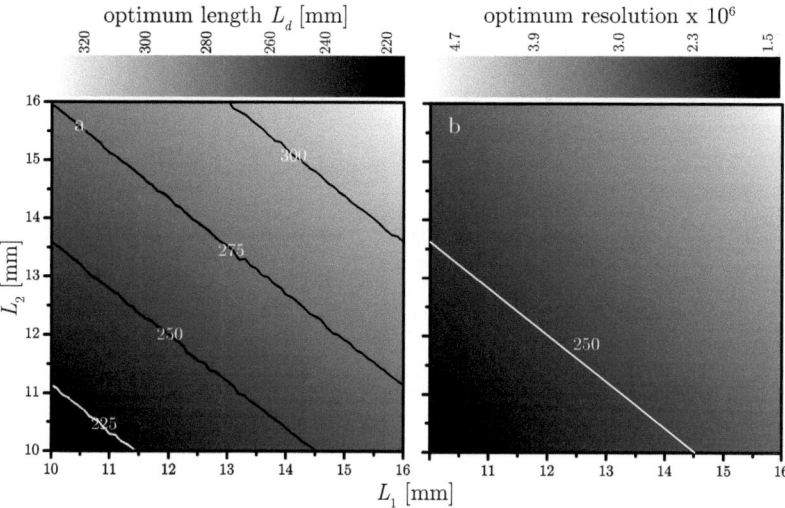

Figure 4.5 Optimization with variation of the accelerator length L_1 and L_2 where L_3 is given by $L_3 = 50$ mm $-(L_1 + L_2)$. It is assumed that the ions start with a Maxwellian velocity distribution according to a beam temperature of $T_\perp = 1$ K along the beam diameter of 3 mm at 6 kV acceleration. **a)** Optimum distance to the space focus plane L_d **b)** Resulting optimum resolution for the parameter sets of L_1, L_2, L_3 and the optimized distance L_d shown in a).

rameter space used for the calculation (L_1 and L_2 between 10 mm up to 16 mm) and only show the optimization trends. Summarizing these results it was shown that a shorter three stage accelerator can fulfil the geometrical design consideration of $L_{\text{accel}} = 50$ mm and $L_d = 250$ mm with improving resolution R compared to the two stage Wiley-McLaren configuration. Additionally, shorter acceleration stages can be constructed with better field homogeneity than longer acceleration stages. Here the limit is given by the higher field strength (sparkovers and mesh flexing). Another question which must be answered is the sensitivity of the configuration regarding errors in the design and the acceleration voltages. In figure (4.6) the two stage Wiley-McLaren configuration and the three stage accelerator configuration are compared. It can be seen that the three stage configuration is more tolerant towards design errors than the Wiley-McLaren configuration *if the acceleration voltages are also optimized*. For small changes in the space focus distance the resolution of the Wiley-McLaren configuration decreases nearly one order of magnitude even if the voltage is optimized. Contrary to this the resolution of the three stage configuration does not noticeably change for changes in L_d

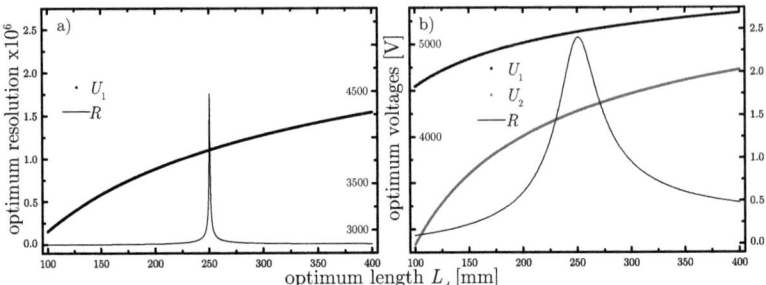

Figure 4.6 Comparison of the two stage Wiley-McLaren configuration (left) with a three stage accelerator (right). Both configurations are optimized for a space focus plane distance of $L_d = 250$ mm. The two stage Wiley-McLaren configuration is more sensitive to deviations in L_d than the three stage design. With the three stage design deviations in L_d can be corrected by optimization of the voltages more effectively than the two stage design. In both calculations it is assumed that the ions start with a Maxwellian velocity distribution according to a beam temperature of $T_\perp = 1$ K along the beam diameter of 3 mm at 6 kV acceleration. **a)** Standard Wiley-McLaren configuration with $L_1 = 36$ mm and $L_2 = 40$ mm. **b)** Three stage design with $L_1 = 12$ mm, $L_2 = 12$ mm and $L_3 = 26.5$ mm.

around several millimeters (see figure 4.6 b). For the case that the voltages are not optimized for the deviation of L_d the three stage design is more sensitive than the Wiley-McLaren configuration. With increasing number of acceleration stages the number of possible error sources increases (design tolerances, power supply voltage stability and precision). Thus for a four stage accelerator at least nine error sources exist (five length and four potentials) limiting the benefit of additional stages. Additionally more acceleration stages mean more meshes which reduce the overall transmission of the device. In that sense the three stage configuration seems to be a promising choice.

4.1.2 The TOFMS-Accelerator: Simulation and Design

According to the results of numerical optimization it was decided to build a three stage accelerator. However, in the numerical optimization calculations an "ideal" accelerator with perfectly homogeneous acceleration fields was assumed. Additionally it was assumed for simplicity that the ion motion and distribution is limited to one dimension. In that case every ion senses the same electric field and thus electric field strength. However, in "real" configurations boundary conditions affect electric field homogeneity and decrease resolution. Hence comprehensive simulations are demanded to construct acceleration stages which provide nearly homogeneous electric field distributions along the acceleration path. Therefore accelerator stage designs were simulated to minimize electric field distortions. For the simulation of different acceleration geometries and setups the ion and electron optics simulation software SIMION 3D version 7 was used [271–273]. Different acceleration geometries were simulated by analyzing the potential gradient and letting ions fly through the accelerator. For the simulations at first a potential array is defined where the accelerator geometries are drawn (electrode definition with geometry files is also possible, see appendix A.2). In the following SIMION calculates by a refine process based on the finite difference method the electric field by solving the boundary value problem's Laplace equation. After the refining process different ion groups can be defined and flown as well the equipotential lines (or field gradients) can be displayed. Different values of the ion and ion-trajectories e. g. TOF can be recorded in a file to determine the resolution of the system. To compare the results obtained by the numerical optimization with the SIMION simulations, the same amount of ions was simulated in SIMION and for the numerical optimization calculations (1001 ions). Additionally the resulting resolution was calculated in a similar way done before by the numerical optimization method including the spatial distribution probability (with and without velocity distribution, see 2.2.5). In that case it was possible to check the integrity of the results obtained by numerical optimization with the SIMION simulations. Here again at first the ion motion was limited to one dimensional motion (no transversal velocity component $v_\perp$). To probe the influence of boundary effects and thus the field homogeneity the ions were started equidistantly arranged in three different lines (ion packages) around the center of the accelerator (accelerator center $y = 0$ mm and due to rotationally symmetric accelerator configuration $\pm y = 10$ mm or $\pm y = 15$ mm see figure 4.7). Where the line width is defined by the beam width (e. g. 3 mm) given by the skimmer diameter of 3 mm. The simulated "ideal" accelerator consisting of four meshes is depicted in figure (4.7). The mesh spacing is $L_1 = 12$ mm, $L_2 = 12$ mm and $L_3 = 25$ mm. The depicted configuration has a diameter of $\varnothing = 76$ mm. Here the meshes are drawn up to the boundaries of the potential array whereas SIMION assumes that the meshes are infinitely long. Therefore the equipotential lines in the acceleration stages are ideally parallel indicating maximum field homogeneity. Thus the resolution

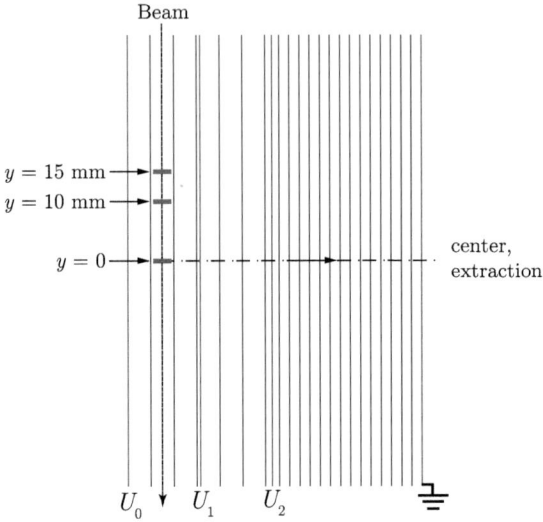

Figure 4.7 The "ideal" three stage accelerator configuration with $L_1 = 12$ mm, $L_2 = 12$ mm, $L_3 = 25$ mm and $\varnothing = 76$ mm. Optimized voltages from the numerical calculations are used ($U_0 = 6$ kV, $U_1 = 5095.79$ V, $U_2 = 4201.87$ V and $L_d = 228.5$ mm). Displayed are the equipotential lines (red lines). SIMION assumes here infinitely long meshes resulting in maximum field homogeneity. With such a configuration the resolution obtained by numerical optimization calculations was reproducible ($R = 2.57 \times 10^6$).

obtained with this setup did not depend on the starting position relative to the middle of the accelerator. The obtained resolution with this setup is equal to the resolution obtained by numerical optimization ($R = 2.57 \times 10^6$). However the situation dramatically changes when a shielding was included and the meshes did not end at the boundaries of the potential array. Such an accelerator configuration is depicted in figure (4.8). Consequently the resulting configuration exhibits then a reduced field homogeneity induced by the boundary conditions (imposed by the shielding). In that case the resolution also is reduced dramatically and was determined for the ion package starting in the center of the accelerator ($y = 0$ mm $\rightarrow R = 9381$) and 10 mm above from the center of the accelerator ($y = 10$ mm $\rightarrow R = 2534$). This effect can be reduced by adding additional ring electrodes between the meshes which stepwise adapt the potential between the two meshes. In that case a resistor chain can be used to adjust the potential of the additional electrodes whereas pulsed operation of such systems is avoided due to the slow time response. Best results for multiple thin electrodes were found for configurations

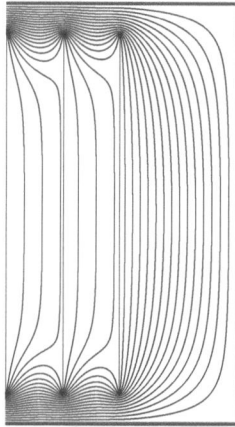

Figure 4.8 A "real" accelerator configuration with $L_1 = 12$ mm, $L_2 = 12$ mm, $L_3 = 25$ mm and $\varnothing = 76$ mm with a shielding around the meshes. Optimized voltages from the numerical calculations are used ($U_0 = 6$ kV, $U_1 = 5095.79$ V, $U_2 = 4201.87$ V and $L_d = 228.5$ mm). Displayed are the equipotential lines (red lines). The boundary condition imposed by the shielding reduces field homogeneity inside the acceleration stages. Hence the resolution decreases by many orders of magnitude to several thousands in contrast to the "ideal" configuration displayed in the figure (4.7) before.

where the electrode thickness is equal to the spacing between the electrodes (e. g. 0.5 mm electrode thickness to 0.5 mm spacing between the ring electrodes). Depending on the geometry using electrodes with larger thickness and spacing than the 0.5 mm resulted in poorer field homogeneity and thus resolution. In that sense multiple ring electrodes have two main drawbacks. For long acceleration stages, e. g. 26 mm, lots of electrodes (e. g. 26 electrodes for the 0.5 mm configuration) are needed which complicate the construction of such systems (see appendix A, figure A.1 b). Additionally due to the use of a resistor chain these electrodes are not well suited for pulsed operation. The setup of the first stage with multiple ring electrodes must be avoided for orthogonal extraction where the ions are generated outside the acceleration stage and pulsed operation is required. Here the question arises if it is possible to find an electrode configuration suited for pulsed operation which provides improved field homogeneity as well ease of construction. Solution for this problem was available in form of electrodes in "pot-shape" [274]. Here two "pot-shaped" electrodes are used instead of multiple ring electrodes between the two meshes of an acceleration stage. These electrodes shield the "inner" side of the accelerator field against perturbations from the outside and achieve higher field homogeneity inside the acceleration stage. An accelerator consisting of two of such "pot-shaped" electrodes and one multiple ring electrode system is depicted in figure (4.9). An alternative to the configuration depicted in figure (4.9) would be to use a "pot-shaped" electrode configuration for the third stage, too. This would have the advantage to operate all three stages pulsed. However, in that case the third stage must be shortened to not decrease field homogeneity. For ease of construction the third stage can be made 12 mm long like the two other stages. The resulting configuration of an accelerator with three "pot-shaped" electrodes and equipotential lines is displayed in figure (4.10). For a picture of the con-

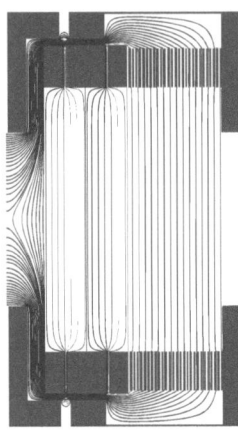

Figure 4.9 A "real" accelerator configuration with $L_1 = 12$ mm, $L_2 = 12$ mm, $L_3 = 26.5$ mm and $\varnothing = 76$ mm with a shielding around the meshes. The first and the second accelerator stage consist of four "pot-shaped" electrodes (1 mm spacing and 5.5 mm thickness) and the third stage consist of 26 ring electrodes (0.5 mm thickness and 0.5 mm spacing). Optimized voltages from the numerical calculations are used ($U_0 = 6$ kV, $U_1 = 5112.94$ V, $U_2 = 4236.92$ V and $L_D = 232.6$ mm). Displayed are the equipotential lines (red lines). The configuration with "pot-shaped" electrodes and ring electrodes shields well the field distortion by boundary conditions imposed by the outer shielding resulting in homogeneous fields inside the acceleration stages. Hence the resolution obtained in the middle of the accelerator is nearly unchanged compared to the calculated optimized resolution (resolution obtained by R: $R = 2.71 \times 10^6$, for $y = 0$ mm $\rightarrow R = 2.69 \times 10^6$, $y = 10$ mm $\rightarrow R = 2.61 \times 10^6$ and $y = 15$ mm $\rightarrow R = 0.83 \times 10^6$).

structed "real" TOFMS-accelerator see appendix A figure (A.1 a). By shortening the length of the third stage the length L_d decrease too ($L_d = 215.7$ mm in contrast to $L_d = 232.6$ mm). Hence the resulting resolution for the configuration in figure (4.10) is lower than the calculated resolution for the configuration in figure (4.9) (resolution obtained by R: $R = 2.04 \times 10^6$ compared to $R = 2.71 \times 10^6$, SIMION: for $y = 0$ mm $\rightarrow R = 2.02 \times 10^6$, $y = 10$ mm $\rightarrow R = 1.97 \times 10^6$ and $y = 15$ mm $\rightarrow R = 0.72 \times 10^6$). However, despite the decrease in theoretical resolution the configuration depicted in figure (4.10) would drastically simplify the construction of the accelerator. Therefore it was decided to investigate the electric field homogeneity inside "pot-shaped" electrodes in dependence of the stage length and diameter. In the figures before electric field homogeneity was displayed by equipotential field lines. In areas of the accelerator where the equipotential field lines are ideally parallel the resulting field is assumed to be homogeneous whereas this is quiet a rough measure for field homogeneity. Physically the electric field is given by the negative gradient of the electric potential ($\mathbf{E} = -\nabla \phi$ where $\phi(x, y, z)$ is the scalar field representing the electric potential at a given point). Therefore the change in the electric field strength ($\Delta \mathbf{E}$) is a well suited measure for field homogeneity. Thus in a domain of the acceleration stage where the value of E ($E = |\mathbf{E}|$) does not change the field in this domain can be regarded as homogeneous. Besides the equipotential lines SIMION can also be used to display the field gradient for given values of the electric field. In that case SIMION displays only changes in the field value above $\Delta E = \pm 10^{-4}$ V/mm assuming that there is no significant change below this value. This assumption of

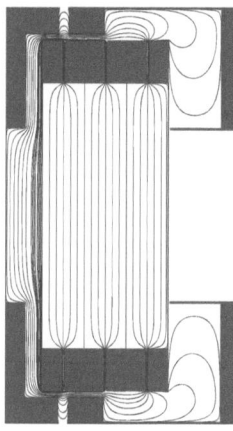

Figure 4.10 A "real" accelerator configuration with $L_1 = 12$ mm, $L_2 = 12$ mm, $L_3 = 12$ mm and $\varnothing = 76$ mm with a shielding around the meshes. All three accelerator stages consist of six "pot-shaped" electrodes (1 mm spacing and 5.5 mm thickness). Optimized voltages from the numerical calculations are used ($U_0 = 6$ kV, $U_1 = 4986.96$ V, $U_2 = 3990.47$ V and $L_D = 215.7$ mm). Displayed are the equipotential lines (red lines). The configuration with "pot-shaped" electrodes shields well the boundary condition imposed by the outer shielding resulting in homogeneous fields inside the acceleration stages. Hence the resolution obtained in the middle of the accelerator is nearly unchanged compared to the calculated optimized resolution (resolution obtained by R: $R = 2.04 \times 10^6$, for $y = 0$ mm $\rightarrow R = 2.02 \times 10^6$, $y = 10$ mm $\rightarrow R = 1.97 \times 10^6$ and $y = 15$ mm $\rightarrow R = 0.72 \times 10^6$).

the SIMION program was taken as homogeneity criterion for the simulated "pot-shaped" electrodes. A potential difference of 1000 V was applied between the two "pot-shaped" electrodes. For a stage length of $L_{pot} = 12$ mm the calculated "ideal" value is $E = 83.3333$ V/mm ($E = 1000$ V/12 mm). In that sense the domain inside the acceleration stage where the field is equal to $E = 83.3333$ V/mm and does not change more than about $\Delta E = \pm 10^{-4}$ V/mm can be regarded as an homogeneous domain. A representative figure of a "pot-shaped" electrode acceleration stage is depicted in figure (4.11). The electric field gradient lines displayed in figure (4.11 a) show a slight increase of the gradient near to the middle ($y = 0$ mm) of the acceleration stage. However, a very steep increase in (ΔE) is observed near the "pot-edge". The behavior of the E-field at half accelerator length ($x = 6$ mm) is displayed in more detail in figure (4.11 b). Up to 8 mm around the middle of the accelerator the E-field can be regarded as homogeneous ($E = 83.3333$ V/mm, $\Delta E = \pm 10^{-4}$ V/mm) and above 8 mm up to 18 mm there is just a slight increase in E ($E = 83.33$ V/mm, $\Delta E = \pm 10^{-2}$ V/mm) whereas the increase in E above this value is very steep and exponential. The gradient field of "pot-shaped" electrodes (see figure 4.11 a) shows two main features. An elliptic shaped domain in the middle of the accelerator and two half-elliptic shaped domains symmetric on the edges near the meshes. Both domains form the border between the homogeneous area and the area with changes in the E-field. In the following many "pot-shape" acceleration stage configurations were simulated and these features displayed. The diameter of the homogeneous domain was determined for both features (middle and edge). Therefore acceleration stages with different inner diameters ($\varnothing = 60$ mm–$\varnothing = 140$ mm) and different stage lengths

4. CHAPTER — 4.1 TOFMS Optimization

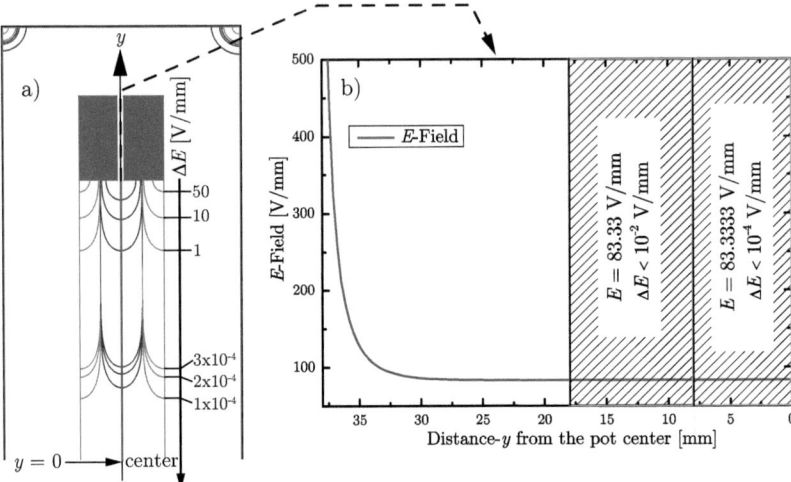

Figure 4.11 An acceleration stage configuration with "pot-shaped" electrodes $L_{pot} = 12$ mm, and $\varnothing = 76$ mm with a shielding around the meshes. The acceleration stage consist of two "pot-shaped" electrodes (1 mm spacing and 5.5 mm thickness). Potential difference between the electrodes is 1000 V ($E = 83.3333$ V/mm in the middle $y = 0$ mm). **a)** Displayed are the gradient lines (green and blue) for changes in E ($\Delta E = \pm 10^{-4}$ V/mm–$\Delta E = \pm 50$ V/mm). **b)** E-field strength along the half acceleration stage length from the "pot-edge" ($y = 38$ mm) up to the middle of the acceleration stage ($y = 0$ mm). The hatched area up to 8 mm shows the domain with a homogeneous E-field ($E = 83.3333$ V/mm, $\Delta E = \pm 10^{-4}$ V/mm) and the hatched area between 8 mm up to 18 mm shows the domain where the increase in the E-field is very low ($E = 83.33$ V/mm, $\Delta E = \pm 10^{-2}$ V/mm).

($L_{pot} = 8$ mm–18 mm) were simulated. Generally the homogeneous domain diameter increases nearly linearly (slope = 1) with increasing inner pot diameter. This behavior is depicted in the figure (4.12 a) for the middle and edge gradient field features. Contrary to this the "pot-shape" acceleration stage length L_{pot} shows the opposite behavior (a negative slope). The diameter of the domain with homogeneous field is inversely proportional to the length L_{pot}. The diameter of the homogeneous domain decreases stronger with increasing L_{pot} (see figure 4.12 b). Hence, a rough linear fit shows a decrease of a factor of 5 with increasing length L_{pot}. These results and the results obtained for different acceleration stage configurations show that "pot-shaped" electrodes are a really good alternative for ring shaped acceleration stages. Best results can be obtained for short acceleration length and for large pot inner diameters. Here the limit is given by the geometrical requirements and mesh flexing at large mesh diameters and acceler-

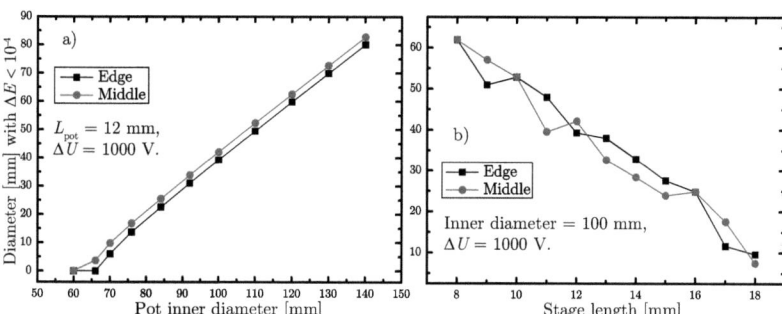

Figure 4.12 Behavior of the homogeneous domain ($\Delta E = \pm 10^{-4}$ V/mm) simulated for different pot shaped acceleration stages. Potential difference between the electrodes is 1000 V ($E = 83.3333$ V/mm in the middle $y = 0$ mm). **a)** Diameter of the homogeneous domain in dependence of the inner pot diameter. Acceleration stage configurations with "pot-shaped" electrodes $L_{pot} = 12$ mm, and different pot inner diameter ($\varnothing = 60$ mm–140 mm). Determined for both gradient features on the edges of the stage and in the center of the acceleration stage (see figure 4.11 a)). The acceleration stage consists of two "pot-shaped" electrodes (1 mm spacing and 5.5 mm thickness). The homogeneous domain diameter increases nearly linearly with increasing pot inner diameter. **b)** Diameter of the homogeneous domain in dependence of the acceleration stage length L_{pot}. L_{pot} was changed between 8 mm up to 18 mm. The pot inner diameter remained constant at 100 mm (electrode spacing constant at 1 mm). With increasing acceleration stage length L_{pot} the diameter of the homogeneous domain decreases.

ation potentials. Besides these aspects an additional advantage of "pot-shaped" electrodes is that these electrode configurations are well suited for pulsed operation (no resistor chain, low capacity and thus fast response time). Moreover by the use of "pot-shaped" electrodes the performance of the TOFMS accelerator can be improved with simplifying the design simultaneously.

4.1.3 The Deflector

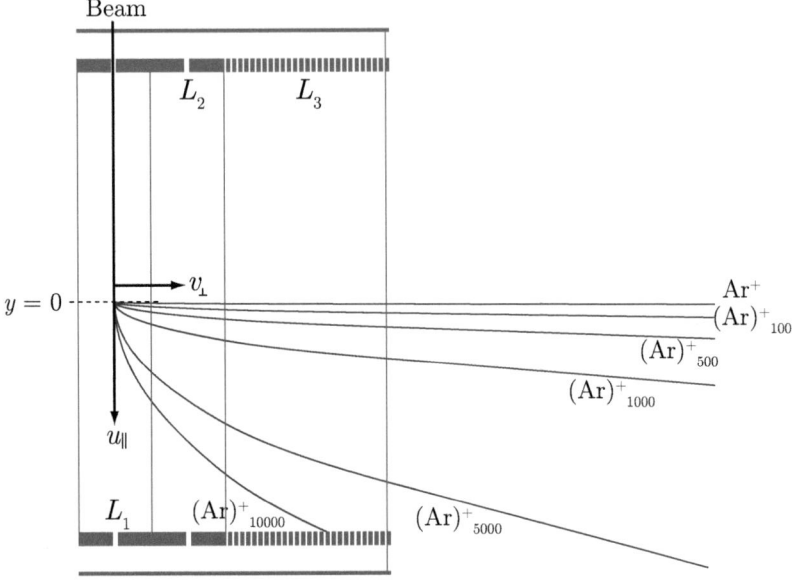

Figure 4.13 Orthogonal extraction of Ar^+-ions with a three stage accelerator ($L_1 = 12$ mm, $L_2 = 12$ mm, $L_3 = 26.5$ mm and $\varnothing = 76$ mm, similar configuration as in figure 4.9). The ions enter the accelerator with an assumed molecular beam velocity of $\langle u_\| \rangle = 630$ m/s (estimated for Ar with equation (2.5) for $T_0 = 380$ K) and are orthogonally extracted by an 6 kV extraction voltage (repeller). It is observed that very heavy clusters (e.g. Ar^+_{10000}) are too slow to leave the accelerator. The other big clusters require deflection plates to compensate the transversal beam velocity component $\langle u_\| \rangle$.

After the optimization of the accelerator the optimization and simulation of the deflector was required. A simple basic deflector consists of two deflection plates parallel to each other. One plate can be grounded and a potential can be applied to the opposite plate (unipolar operation). When two potentials with opposite polarization are applied to the plates it is referred to as bipolar operation. The deflector is needed for the compensation of the beam parallel velocity component $\langle u_\| \rangle$ given by the velocity distribution of the molecular beam (see 2.1.1). Except the velocity slip every ion in the molecular beam gains this velocity during the expansion (narrow velocity distribution). This velocity component defines the kinetic energy of the ions perpendicular to the extraction direction in orthogonal extraction mode operated TOFMS. Due to the dependence of the kinetic energy on the velocity and mass of the ions this kinetic energy component is negligible for small cluster ions extracted by strong accelerator extraction fields (several kV). However, with increasing cluster size the kinetic energy in beam translational direction increases too (see 2.1.3) and makes a compensation inevitable. The cluster ions enter the first accelerator stage perpendicular to the extraction direction. A voltage pulse is applied to the accelerator to extract the ions perpendicular to the translational direction of the molecular beam (orthogonal extraction mode). Here it depends on the ratio between the beam velocity component $\langle u_\parallel \rangle$ (given by the molecular beam properties) to the extraction velocity component $v_\perp$ (mass dependent, given by the extraction potential) whether the ion can leave the accelerator or not. This problematic situation for very big clusters is shown in figure (4.13). Depicted are the different simulated ion trajectories for argon clusters with different sizes (Ar^+, Ar_{100}^+, Ar_{500}^+, Ar_{1000}^+, Ar_{5000}^+ and Ar_{10000}^+). The clusters enter with the same assumed transversal beam velocity of ($\langle u_\parallel \rangle$ = 630 m/s) the accelerator (estimated with equation 2.5 for Ar and T_0 = 380 K). With the application of the 6 kV extraction potential the ions in the middle of the accelerator (highest field homogeneity) are accelerated perpendicular to the transversal beam velocity $\langle u_\parallel \rangle$. By this acceleration the ions gain an additional extraction velocity component ($v_\perp$). In dependence of the mass and thus the ratio between $\langle u_\parallel \rangle$ and $v_\perp$ the ions follow different trajectories after extraction. As shown in figure (4.13) it would be very hard to detect ions with masses above 200000 amu (Ar_{5000}^+) with orthogonal extraction at 6 kV. In the case when a very heavy ion can leave the accelerator a strong deflection field is required to compensate the transversal component v_{trans}. The mass rage of transmitted heavy ions can be enhanced by extracting the ions near the beam entrance of the accelerator (lower field homogeneity and thus resolution). The geometry of the deflector directly affects the two relevant parameters:

Transmission (Intensity) The lengths of the deflector plates influence the number of transmitted heavy ions (simulated with Ar_{1000}^+ ions). Longer plates require lower potentials for deflection (increased ion fly through times) and do not decrease significantly resolution. Contrary longer deflection plates reduce the number of transmitted ions (ions that can leave the deflector without colliding with the plates). Depending on deflector geometry (plate length, distance and width) and field homogeneity between the deflection plates the beam shape can change. Generally a lens effect for inhomogeneous fields which widens the beam was observed. Therefore a part of the beam can miss the detector reducing the detectable beam intensity.

Resolution Influence of the deflection process on the overall obtained resolution. Determined by the time-of-flight distribution for a ion matrix package of

Ar_{25}^+ ions at the space focus plane of the optimized three stage accelerator ($L_1 = 12$ mm, $L_2 = 12$ mm and $L_3 = 26.5$ mm).

These two parameters (transmission (intensity) and resolution) are not independent from each other. Therefore one geometry value of the deflector (plate width, plate length, plate distance and position relative to the accelerator) was changed and the impact on these parameters was observed where two geometric measures were hold at fixed values. For the simulations in SIMION the optimized "real" accelerator (see figure 4.9) and optimized reflectron were used (optimization of the reflectron will follow this subsection). The detector is represented by a thin disc with 25 mm diameter. The deflector is placed centered in front of the accelerator within a shielded housing with entrance and exit slits (for optimum shielding of the deflection fields). These components were placed in a SIMION ion-workbench according to the real measures of the apparatus. For the simulations two ion sizes were used (Ar_{1000}^+ for the transmission behavior and Ar_{25}^+ for the resolution determination). The transmission and intensity of the deflector was determined by the heavy species Ar_{1000}^+ arranged diagonally in a 20 mm long ion package with 2 mm width (101 ions, see e. g. figure 4.14 a) the accelerator and ion trajectories). These ions were started near the beam entrance of the accelerator. The "transmission" performance of the deflector geometry was assessed by the number of ions which can leave the deflector. The "intensity" performance of the deflector geometry was assessed by the number of ions which reach the detector. In the case of the "resolution" performance the lighter ionic species Ar_{25}^+ were simulated. Contrary to the heavy Ar_{1000}^+ ion group the Ar_{25}^+ were started around the middle of the accelerator (best field homogeneity, see figure 4.19 a). The Ar_{25}^+-ion package consisted of a matrix of ions arranged in equidistantly distributed 11 lines (width = 2 mm, 101 ions in each line) forming a ion package length of 20 mm (altogether 1111 ions). For the accelerator and the reflectron the optimized voltages calculated by numerical optimizations were used. In the case of the deflector the potentials were adjusted for optimum beam intensity by repeated simulations for different deflection potential values (maximum transmission and counts on the detector). Best resolution performance results were obtained for the bipolar operation of the plates (same potential value with opposite polarity). The time-of-flights of the Ar_{25}^+ ions to the space focus plane was recorded and the resolution was determined by the TOF distribution. Additional shielding plates (grounded) in front and after the deflection plates were used for better field homogeneity. The first value that was varied was the distance between the deflection plates. Without changing the other values the distance between the deflection plates was varied between 40 mm up to 80 mm. For large plate distances e. g. 80 mm a lens effect was observed. The deflector focused at first the ion package into the reflector whereas the reflected beam reached the detector with a much wider shape than usual. This behavior for the Ar_{1000}^+ ions is depicted in figure (4.14). Depicted are the ion trajectories of the Ar_{1000}^+ ions (blue) and cut-through views

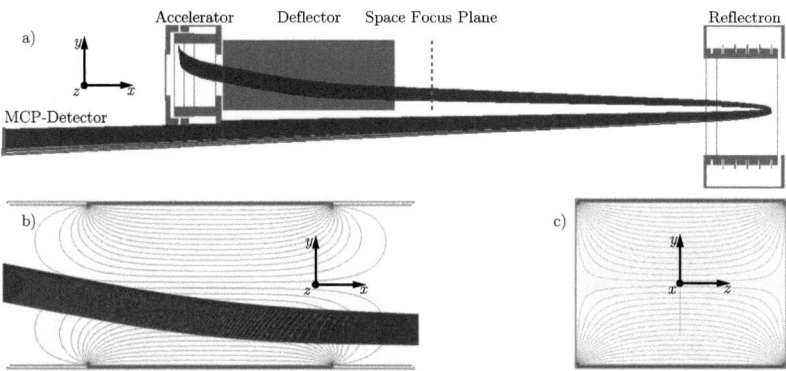

Figure 4.14 SIMION simulation of the Re-TOFMS setup for the orthogonal extraction of heavy cluster ions (Ar_{1000}^+). Optimized geometries of the accelerator and reflectron are used with numerically optimized potential values ($R = 1.17 \times 10^6$ obtained for the deflected Ar_{25}^+ ions, $R = 2.6 \times 10^6$ without deflection, see subsection 4.1.2). **a)** Depicted are the ion trajectories of Ar_{1000}^+ ions (blue lines) deflected by a deflector with a distance between the deflection plates of 80 mm (bipolar ±605 V, optimum value adjusted by repeated simulations). Beam widening at the detector "disc" is observable. **b)** Cut through view (in beam direction, xy-plane) of the deflector with ion trajectories and equipotential lines (red lines). One shielding plate in front and one shielding plate after the deflection plates are visible (shielding plate lengths are each 40 mm, deflection plate lengths are each 120 mm, all plates have a width of 100 mm and the overall deflector length is 200 mm). **c)** Cut through view (yz-plane) perpendicular to the beam direction in the center of the deflector with equipotential lines (red).

of the deflector with equipotential lines (red line b) and c) in figure 4.14). In the case when the distance between the deflection plates is too small e. g. 45 mm (see figure 4.15) a portion of the ions cannot leave the deflector. Thus the beam transmission decreases and the number of ions that reach the detector, too. Regarding the simulated resolving power, the resolution of the whole Re-TOFMS apparatus increases with the distance between the deflection plates. This can be explained by the trajectories of the Ar_{25}^+ ions (see figure 4.16). In the case of small distances the ions fly near the deflection plates and "sense" more differences of the potential field. Thus the individual time of flight between the ions differ more and reduce resolution. The ion matrix of Ar_{25}^+ ions used in the simulations is 20 mm long and the distance between the deflection plates e. g. 45 mm. The ion matrix is than distributed over the half distance of the deflection plates. In the case of larger distances between the deflection plates this ratio changes to 1/4. This behavior is depicted in the figure (4.16). It can be seen in figure (4.16) that

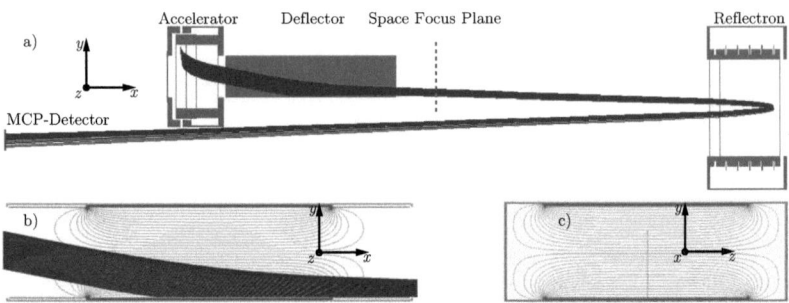

Figure 4.15 SIMION simulation of the Re-TOFMS setup for the orthogonal extraction of heavy cluster ions (Ar_{1000}^+). Optimized geometries of the accelerator and reflectron are used with numerically optimized potential values ($R = 0.14 \times 10^6$ obtained for the deflected Ar_{25}^+ ions). **a)** Depicted are the ion trajectories of Ar_{1000}^+ ions (blue lines) deflected by a deflector with a distance between the deflection plates of 45 mm (bipolar ±333 V, optimum value adjusted by repeated simulations). Due to the lower distance between the deflection plates a potion of the ions collide with the plates and cannot leave the deflector. **b)** Cut through view (in beam direction, xy-plane) of the deflector with ion trajectories and equipotential lines (red lines). One shielding plate in front and one shielding plate after the deflection plates are visible (shielding plate lengths are each 40 mm, deflection plate lengths are each 120 mm, all plates have a width of 100 mm and the overall deflector length is 200 mm). **c)** Cut through view (yz-plane) perpendicular to the beam direction in the center of the deflector with equipotential lines (red).

the trajectories of the Ar_{25}^+ ions for the case of a small distance between the plates pass a more bent potential energy surface than in the case of the deflection plates with a larger distance. The ion trajectories seem to be more curved which result in greater differences in the TOF for the ions which fly near the deflection plates. Finally these differences in the potential energy surface curvature result in greater TOF distributions and thus lower resolution. However, in our case it was more important to obtain a better intensity (collimated beam) than obtaining the best theoretical resolution. Therefore it was decided to use smaller plate distances like e. g. 50 mm than 80 mm where the beam on the detector is widened and intensity is reduced. The next geometrical parameter of the deflector which was changed was the length of the deflection and shielding plates. The other parameters were hold at fixed values and the length of the deflection plates and the shielding plates were incrementally altered. The deflection plates lengths where varied form 60 mm up to 120 mm and the shielding plates length from 20 mm up to 80 mm (each shielding plate, front and rear). However, the total length of the deflector was limited to 200 mm due to the spatial focus plane at $L_F \approx 250$ mm (the position of the planed mass gate). Like in the case of the distance between

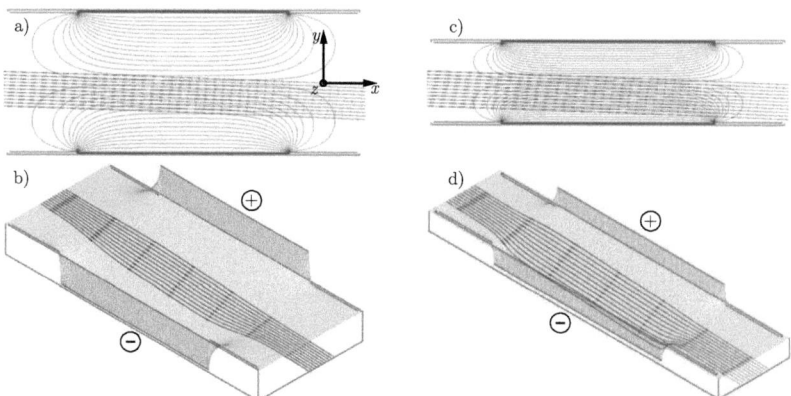

Figure 4.16 Comparison of the SIMION simulations of the two deflector geometries with different distances between the deflection plates (from figure 4.14 and figure 4.15). Depicted are the ion trajectories of Ar_{25}^+ ions flown through the deflectors. A ion matrix of Ar_{25}^+ with (2×20) mm size and 1111 ions was used to calculate the resolution by the TOF distribution. **a)** Depicted are the equipotential lines induced by the bipolar deflection potentials (red lines) and ion trajectories of Ar_{25}^+ ions (blue lines) deflected by a deflector with a distance between the deflection plates of 80 mm (cut through view in beam direction the xy-plane, plates bipolar at ± 92 V, optimum value adjusted by repeated simulations). **b)** Potential energy surface view of the same deflector in a) with ion paths in the potential energy surface. **c)** Depicted are the equipotential lines induced by the bipolar deflection potentials (red lines) and ion trajectories of Ar_{25}^+ ions (blue lines) deflected by a deflector with a distance between the deflection plates of 45 mm (cut through view in beam direction the xy-plane, plates bipolar at ± 48 V, optimum value adjusted by repeated simulations). **d)** Potential energy surface view of the same deflector in c) with ion paths in the potential energy surface.

the deflection plates again many different geometries were simulated. It was observed that longer deflection plates reduce the lens effect which widens the beam at the detector (see e. g. figure 4.14). In contrast the transmission of the deflector for the high mass range simulated by Ar_{1000}^+ ions decreases when the distance between the plates is to low (e. g. 50 mm). Here we will pick out some representative results for different plate lengths and their influence on transmission, intensity and resolution (for plate distances of 50 mm). In the case where the plate lengths are to short a lens effect similar to the effect depicted in figure (4.14) was observed. Such a short deflector with short deflection plates is shown in figure (4.17). Here the deflection plates are 60 mm long and the front and rear shielding plates are 20 mm long. The simulated ion beam of Ar_{1000}^+ is focused into the de-

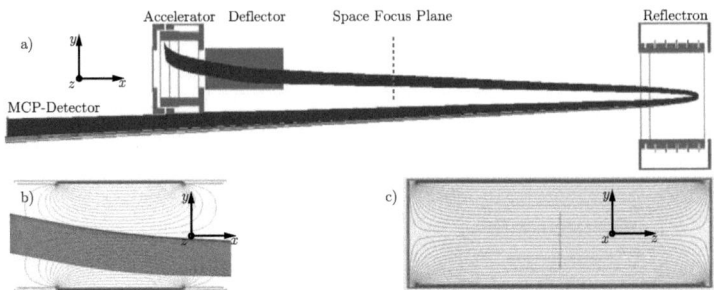

Figure 4.17 SIMION simulation of the Re-TOFMS setup for the orthogonal extraction of heavy cluster ions (Ar^+_{1000}). Optimized geometries of the accelerator and reflectron are used with numerically optimized potential values ($R = 0.27 \times 10^6$ obtained for Ar^+_{25} ions). **a)** Depicted are the ion trajectories of Ar^+_{1000} ions (blue lines) deflected by a deflector with a distance between the deflection plates of 50 mm (bipolar ±710 V, optimum value adjusted by repeated simulations). Due to the shorter plates (deflection plates lengths 60 mm, and shielding plates lengths 20 mm each) beam widening at the detector "disc" is observable. **b)** Cut through view (in beam direction the xy-plane) of the deflector with ion trajectories (Ar^+_{1000}) and equipotential lines (red lines). All plates have a width of 140 mm and the overall deflector length is 100 mm. The plate width of 140 mm is not sufficient to compensate the influence of the shorter plate length **c)** Cut through view perpendicular to the beam direction (yz-plane) in the center of the deflector with equipotential lines (red). A relatively homogeneous field distribution in the center is available which can be attributed to the plate width of 140 mm.

flector due to the lens effect. The reflected ions form a beam which is widened at the detector plane (reduced the simulated intensity). This effect can be reduced by using a longer deflector (longer deflection and shielding plates). But for the case when the whole deflector is too long (longer than 160 mm) the transmission for the heavy ions (Ar^+_{1000}) decreases. The heavy ions cannot leave the deflector and collide with the plates. This fact leads to the limitation of the whole detector length to 160 mm. Here the question arises how long the deflection plates and shielding plates must be. In the simulations before shielding plates length below 20 mm drastically reduced resolution whereas too long shielding plates limit the length of the deflection plates which decrease transmission. However, transmission can be increased by a short shielding plate followed by a long deflection and a longer rear shielding plate. Therefore the first shielding plate can be set to the lowest length of 20 mm and the rest length of 140 mm can be reserved for the deflection plates and rear shielding plates. After the simulation of different ratios between the deflection plates and the rear shielding plates an optimum was find with high transmission (100%) and low beam distortion. Due to the use of a

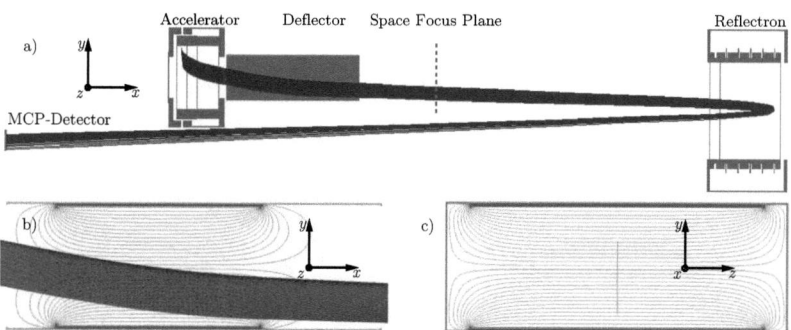

Figure 4.18 SIMION simulation of the Re-TOFMS setup for the orthogonal extraction of heavy cluster ions (Ar_{1000}^+). Optimized geometries of the accelerator and reflectron are used with numerically optimized potential values ($R = 0.33 \times 10^6$ obtained for Ar_{25}^+ ions see figure 4.19). **a)** Depicted are the ion trajectories of Ar_{1000}^+ ions (blue lines) deflected by a deflector with a distance between the deflection plates of 50 mm (bipolar ±498 V, optimum value adjusted by repeated simulations). Due to the optimized geometry all ions are transmitted by the deflector and minimal beam widening at the detector is obtained. **b)** Cut through view (in beam direction the xy-plane) of the deflector with ion trajectories (Ar_{1000}^+) and equipotential lines (red lines). Instead of shielding plates "free room" in front (20 mm) and after (50 mm) the deflection plates (85 mm long) is visible. All plates have a width of 120 mm and the overall deflector length is 155 mm. The plate width of 120 mm is limited by the dimensions of the vacuum chamber. **c)** Cut through view perpendicular to the beam direction (yz-plane) in the center of the deflector with equipotential lines (red). A relatively homogeneous field distribution in the center is available which can be attributed to the plate width of 120 mm and the low distance between the plates of 50 mm.

shielding box around the deflector the additional shielding plates can be replaced only by "free rooms" in front and after the deflection plates. The final deflector obtained by this optimization is depicted in figure (4.18). It consists of deflection plates with 85 mm lengths and a free room in front of the deflection plates with 20 mm lengths and a free room after the deflection plates of 50 mm. The whole resulting deflector is 155 mm long inside a shielding housing with to slits. The width of the plates increases field homogeneity and was limited only by the dimensions of the vacuum chamber to the maximum 120 mm. In figure (4.18 c) the equipotential lines in the middle of the plates are nearly parallel and show that a width of 120 mm is sufficient to obtain homogeneous fields. With this construction all simulated Ar_{1000}^+ ions can leave the deflector without collisions and every ion can reach the detector. Additionally the deflection plates and rear shielding "free room" is long enough to deflect all ions to the detector plane without or

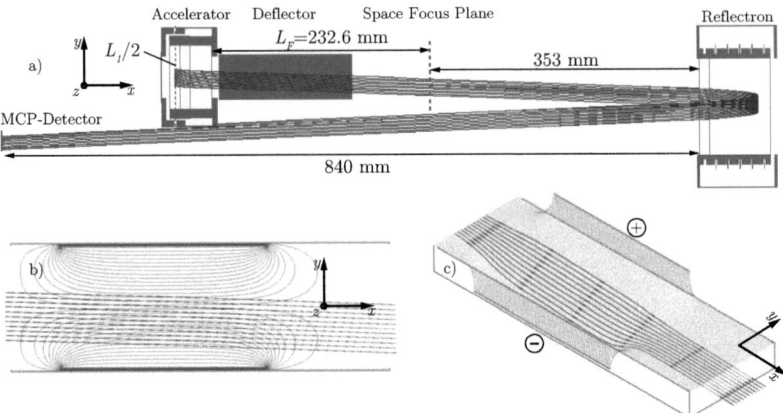

Figure 4.19 SIMION simulation of the Re-TOFMS setup for the orthogonal extraction of light cluster ions (Ar_{25}^+) with the same optimized deflector geometry as in figure (4.18) **a)** Depicted are the ion trajectories of Ar_{25}^+ ions (blue lines) used for the calculation of the theoretical resolution ($R = 0.33 \times 10^6$, distance between the plates 50 mm, bipolar ±30 V, optimum value adjusted by repeated simulations). Due to the optimized geometry all ions are transmitted by the deflector and minimal beam widening at the detector is obtained. **b)** Cut through view (in beam direction the xy-plane) of the deflector with ion trajectories (Ar_{25}^+) and equipotential lines (red lines). Instead of shielding plates "free room" in front and after the deflection plates is visible. **c** Potential energy surface view of the deflector with ion trajectories for the Ar_{25}^+ ion matrix.

minimal beam widening and distortion. This behavior is shown in figure (4.18 a and b) for the simulated ion trajectories of Ar_{1000}^+ ions. The obtained resolution ($R = 0.33 \times 10^6$ determined for Ar_{25}^+) is the maximum possible resolution with maximum transmission (100%) of Ar_{1000}^+ ions and lowest beam widening. The simulation of Ar_{25}^+ ion trajectories ((2×20) mm ion matrix) with equipotential lines and a potential energy surface view of the same setup are depicted in the figure (4.19).

4.1.4 The Mass Gate

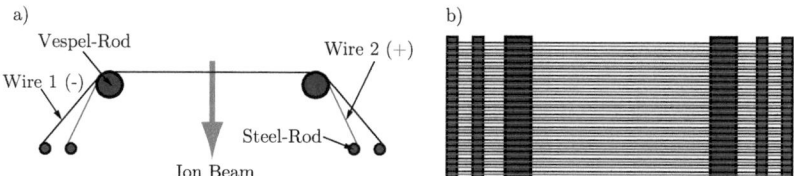

Figure 4.20 Schematic views of the interleaved comb mass gate constructed for size selection of cluster ions. Two coplanar sets of stainless-steel wires tightly strung around rods comparable to a weaving loom. Due to the use of vespel rods and holders the two wires are insulated from each other and can be kept at different potentials. **a)** Schematic side view of the mass gate configuration. **b)** Schematic top view of the mass gate configuration.

One key feature of the experimental setup is the improved design of a pulsed ion gate (9 in figure 3.1), located in the first focal plane of the Re-TOFMS. The design of the mass gate is based on the original ideas of Cravath, Bradbury and Nielsen [275; 276] and enables the size selection of cluster ions prior to their surface interaction. It consists of two parallel coplanar sets of stainless-steel wires in a UHV-compatible frame, offering low capacity and allowing fast switching times (see figure 4.20)[1]. The wire spacing of 500 μm (wire diameter $\varnothing = 50\mu m$) results in an optical transmission of 90%, the mass selection performance exceeds 190 at comparatively small deflection voltages of 150 − 500 V depending on the Re-TOFMS acceleration voltage (see subsection 4.2.4). The main advantage of the interleaved comb mass gate is its steep potential gradient (see figure 4.21 b). Due to the potential compensation of the alternating wire configuration, the potential declines exponentially to zero (see figure 4.21 a and b). Thus, unwanted field perturbations of the field free drift region of the mass spectrometer are minimized simultaneously improving resolution. In operation a bipolar potential is applied to the wires of the mass gate (as in figure 4.21 a). Every ion package which enters the mass gate is deflected by the wire potentials and leaves the normal beam trajectory. These ions collide with the exit slit of the mass gate or with the slit located in front of the reflectron and are "filtered" from the mass spectra. For "gating" the desired ion mass (or cluster size) the potentials applied to the mass gate are switched off by fast push-pull switches (see section 3.4). These ions can pass the mass gate and fly on the normal ion trajectories to the detector. By switching the potentials to the wires on again, following ion packages are also "filtered" from the beam. Thus the size selection performance of the mass gate depends on the resolution at the wire position plane and the switching speed of

---
[1]Designed and constructed by Ulf Bergmann

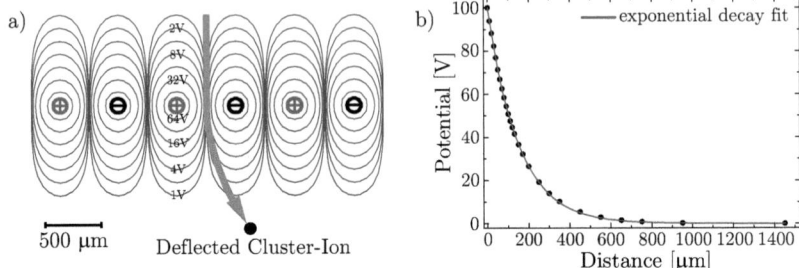

Figure 4.21 Depicted are SIMION simulations of the interleaved comb mass gate. a) SIMION equipotential view of the alternating wire configuration. An alternating potential of ±100 V is applied to the wires. b) Graph showing the potential values from a) in dependence from the wire distance (in beam direction). The decline of the potential is well fitted by an exponential decay function.

the transistor switches. Ideally the mass gate is positioned in the space focus plane of the TOFMS accelerator for the highest size selection performance. Due to the principle of TOFMS mass resolution depends on many factors e. g. mass, beam diameter, ion package length and so on (see section 2.2.4). To estimate the performance of the mass gate before the implementation in the experimental setup we performed SIMION simulations. The simulation of the accelerator and the deflector were described before (in 4.1.2 and 4.1.3). In the simulations before the aim was to optimize each component of the TOFMS for optimal resolution. In contrast for the mass gate a different approach is made. The question that raised was if it is possible to mass select big cluster sizes e. g. $N \approx (CO_2)_{100}^+$ with the present setup. Therefore the optimized three stage accelerator (see figure 4.9) and the optimized deflector (see figure 4.19) were used to simulate ion TOF distributions of $(CO_2)^+$-cluster ions with N around 100 in the space focus plane (in contrast to the Ar_{25}^+ cluster ions used in the simulations before). Due to the use of much higher masses the kinetic energy in beam direction increases to about 9 eV per cluster ion (for $(CO_2)_{100}^+$ clusters assuming a beam velocity of $\langle u_\parallel \rangle = 630$ m/s as in the case of Ar). In the following simulations the cluster-beam is represented by three ion groups. One ion group consists of 101 ions with the mass of a $(CO_2)_n^+$ clusters of the same size ($N = 99$, $N = 100$ or $N = 101$). The ions are positioned equidistantly in three lines in the first stage of the accelerator, forming one group. The length of the ion group lines defines the beam diameter (experimentally defined by the skimmer diameter). To simulate the beam pulse length (given by the valve opening time), the first and the third ion groups are positioned symmetrically around the second ion package, which is located in the center of the acceleration stage (as in the case of the accelerator simulations see 4.1.2). Most of the simulations were performed with only the two acceleration stages of the three

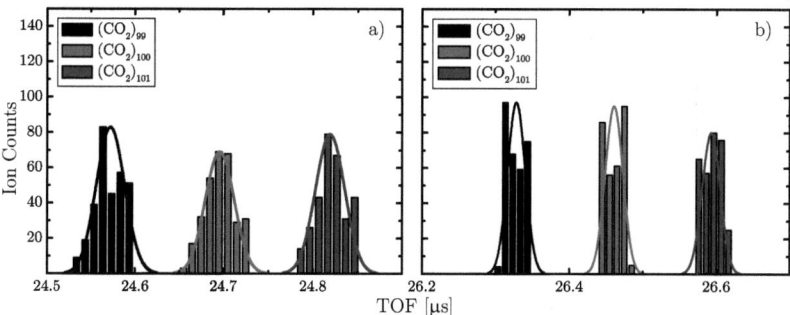

Figure 4.22 Histogram graphs of the TOF distribution for three different cluster sizes ($(CO_2)_{99}^+$, $(CO_2)_{100}^+$ and $(CO_2)_{101}^+$) recorded at the space focus plane of the TOFMS accelerator (wire plane of the mass gate). **a)** TOF distribution for the two stage accelerator ($L_1 = 12$ mm, $L_2 = 12$ mm and $L_D = 326$ mm) at 6 kV acceleration ($U_0 = 6$ kV, $U_1 = 4873.65$ V, ± 95 V for deflection the optimum value adjusted by repeated simulations, 3 mm beam width and 15 mm beam pulse length). The resolution calculated from the TOF spread of each cluster size is $R = 650.1$. **b)** TOF distribution for the three stage accelerator ($L_1 = 12$ mm, $L_2 = 12$ mm, $L_2 = 26.5$ mm and $L_D = 299.5$ mm) at 6 kV acceleration ($U_0 = 6$ kV, $U_1 = 5218.82$ V, $U_2 = 4466.55$ V, ± 95 V for deflection the optimum value adjusted by repeated simulations, 3 mm beam width and 15 mm beam pulse length). The resolution calculated from the TOF spread of each cluster size is $R = 939$.

stage accelerator to simulate a "worst case" scenario ($L_1 = 12$ mm, $L_2 = 12$ mm and the length of $L_3 = 26.5$ mm is added to $L_d = 299.5$ mm + $L_3 = 326$ mm for two stage operation and for three stage operation $L_d = 299.5$ mm). The time-of-flight distributions of the ions were recorded at the space focus plane located at the distance L_d. The resulting resolution was calculated with equation (2.36) by assuming a equipartition distribution for the starting ions ($p(x_s, i) = 1/n$). By displaying the TOF distribution of the three different cluster sizes ($(CO_2)_{99}^+$, $(CO_2)_{100}^+$ and $(CO_2)_{101}^+$) in a histogram graph the TOF difference between the three sizes can be deduced. For successful mass gating the difference in arrival time for each cluster size must be greater than the minimum switching time of the transistor push-pull switches. Two representative histograms for 6 kV acceleration for two stage operation and three stage operation of the TOFMS accelerator are shown in figure (4.22). The resolution obtained by the TOF spread for the two stage configuration shown in figure (4.22 a) is $R = 650.1$ which is about 2/3 of the resolution obtained for the three stage configuration (figure 4.22 b) with $R = 939$. The arrival time difference between the last bin of the $(CO_2)_{99}^+$ cluster ion package and the first bin of the $(CO_2)_{101}^+$ cluster ion package is 180 ns for the two stage configuration (4.22 a) and 220 ns respectively for the three

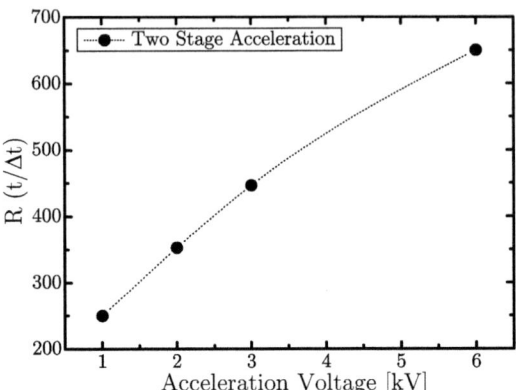

Figure 4.23 Influence of the acceleration voltage on the resolution at the space focus plane (wire plane of the mass gate) obtained from the TOF distribution for three different cluster sizes ($(CO_2)_{99}^+$, $(CO_2)_{100}^+$ and $(CO_2)_{101}^+$). SIMION TOFMS resolution for the two stage accelerator configuration ($L_1 = 12$ mm, $L_2 = 12$ mm and $L_D = 326$ mm) at 1–6 kV acceleration (± 50–95 V for deflection, optimum value adjusted by repeated simulations, 3 mm beam width and 15 mm beam pulse length).

stage configuration (4.22 b). The minimum output pulse width of the transistor switches is 200 ns. Thus it would be barely possible to mass select cluster sizes around $(CO_2)_{100}^+$ with both configurations (two stage and three stage at 6 kV acceleration). For lower acceleration voltages the arrival time difference in the focus plane increases with decreasing acceleration voltage (e. g. 210 ns at 3 kV and 230 ns at 2 kV two stage acceleration). In contrast to this the calculated resolution decreases with decreasing acceleration voltage as might be expected. In this case the ion package peaks gain on width and overlap which each other for acceleration voltages below 2 kV (two stage acceleration). Therefore a clean mass selection for acceleration voltages below 2 kV and cluster sizes around $N = 100$ is not possible due to overlapping peaks even the high resolution ($R = 250$ at 1 kV two stage acceleration). Contrary to the mass resolution definition (see subsection 2.2.4) for "clean" mass selection a high enough mass resolution (narrow peak width) and additionally sufficient arrival time delay between the ion packages of different mass is required (no overlap of mass peaks). For acceleration voltages above 1 kV the arrival time difference between the three cluster sizes is higher than 200 ns and the resolution even for the two stage system is larger than $R = 250$. Thus it can be assumed that for these configurations a mass selection performance above $N = 100$ can be expected. Note here that two factors decide the mass selection performance for a given configuration. The first factor is the time difference between the different cluster size peaks which is given by the acceleration voltage

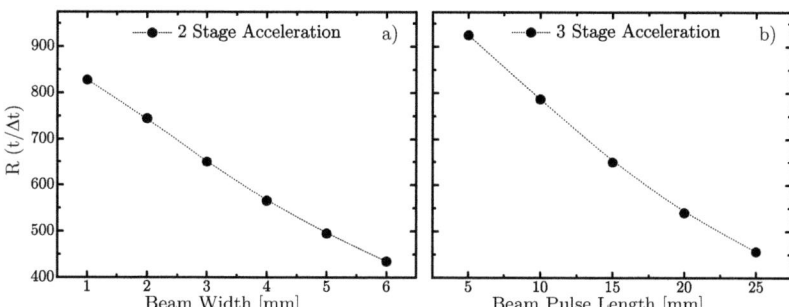

Figure 4.24 Influence of the beam properties (beam width and beam pulse length) on the resolution at the space focus plane (wire plane of the mass gate) obtained from the TOF distribution for three different cluster sizes ($(CO_2)_{99}^+$, $(CO_2)_{100}^+$ and $(CO_2)_{101}^+$). **a)** SIMION TOFMS resolution for the two stage accelerator configuration ($L_1 = 12$ mm, $L_2 = 12$ mm and $L_D = 326$ mm) at 6 kV acceleration (± 95 V for deflection, beam pulse length fixed at 15 mm) and variation of the beam width. **b)** SIMION TOFMS resolution for the three stage accelerator configuration ($L_1 = 12$ mm, $L_2 = 12$ mm, $L_3 = 26.5$ mm and $L_D = 299.5$ mm) at 6 kV acceleration (± 95 V for deflection, optimum value adjusted by repeated simulations beam width fixed at 3 mm) and variation of the beam pulse length.

(for larger acceleration voltages this time difference decreases). The other factor is the available resolution thus the peak width for a given configuration which is narrower for larger acceleration voltages or three stage operation. In the figure (4.23) the dependence of the resolution from the acceleration voltage is depicted. As mentioned above with increasing acceleration voltage the resolution of the system at the space focus plane increases similar to a root function. Besides the acceleration voltage the two main beam properties (beam width and beam pulse length) also affect the resulting resolution. The influence of these two properties on the resulting resolution is depicted in figure (4.24). For comparison in figure (4.24 a) the behavior of the two stage accelerator for increasing beam width and in figure (4.24 b) the behavior of the three stage accelerator for increasing beam pulse length are depicted. In both cases with increasing values for the beam properties the resolution of the system decreases nearly exponentially. These results show that the resolution of the system can be improved by reducing the beam width or beam pulse length. The beam width is given by the skimmer diameter ($\varnothing = 3$ mm) and the orthogonal beam velocity ($v_\perp$). By focusing the beam with an einzel lens into the TOFMS accelerator the beam width can be reduced. However, strong focusing with an einzel lens generates a focal point after which the beam is widened (similar behavior like a focusing optical lens). Therefore a slightly focusing of the beam would be preferred. In both cases a uniform beam width along the whole beam pulse length cannot be achieved. It must be also noted that beam focusing is limited by the coulomb repulsion of the ions limiting the final beam width. Regarding the beam pulse length, this value can be reduced by pulsed ionization of the cluster beam. This can lead to reduced beam intensity which is unwanted for cluster experiments where high beam intensities are required (e. g. scattering experiments). The results of these "worst" case simulations obtained in this subsection show that it must be possible to mass select cluster sizes around $N = 100$ with the present accelerator, deflector and mass gate configuration. This mass selection performance was also approved by experimental results which are summarized in subsection (4.2.4).

4.1.5 The Reflectron and Target Surface

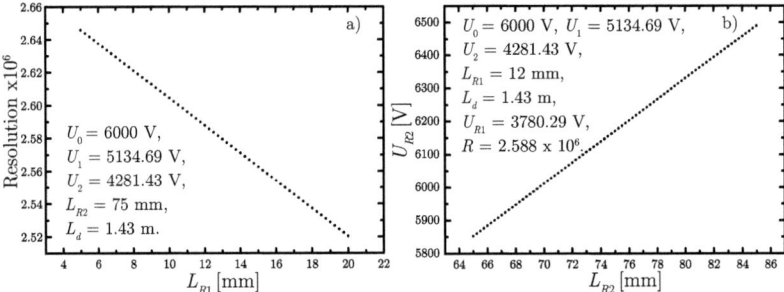

Figure 4.25 Numerical optimization results of the two stage reflectron for one dimensional ion motion. It is assumed that the ions start with a Maxwellian velocity distribution according a beam temperature of 1 K along the beam diameter of 3 mm at 6 kV acceleration. The optimized three stage accelerator geometry and optimized acceleration voltages were used in these calculations. **a)** The second reflectron stage length L_{R2} was hold at a fixed value ($L_{R2} = 75$ mm) and the first stage length L_{R1} was varied. Optimal reflectron voltages U_{R1} and U_{R2} were calculated. The obtained resolution R (•) at the detector plane of the configuration decreases linearly with increasing stage length L_{R1}. **b)** Here the first stage length L_{R1} was hold at a fixed value ($L_{R1} = 12$ mm) and the second reflectron stage length was varied. The obtained resolution of the system is independent from the length L_{R2} of the second reflectron stage. Only the optimal voltage for the second reflectron stage U_{R2} (•) increases linearly with increasing stage length L_{R2}.

The optimization of the reflectron was done analogously to the optimization of the accelerator. The design consideration was to construct two stage reflectron with a desired whole length of about 100 mm and a maximum diameter of 200 mm (designated room in the vacuum chamber). An estimated length of

≈ 15 mm was reserved for the reflectron holder assembly. In that case ≈ 85 mm are available for the reflectron stages which must be arranged in two different stage lengths. Therefore at first numerical optimization calculations with these restrictions were done with R. For the calculations the optimized three stage accelerator geometry ($L_1 = 12$ mm, $L_2 = 12$ mm, $L_3 = 26.5$ mm, spatial focus plane at $L_F = 232.6$ mm see figure 4.9 and 4.19) with optimized voltages was used. The question that arise was how the length of each reflectron stage (L_{R1} and L_{R2}) influences the resolution. Therefore one stage length was hold at a fixed value and the other length was varied and numerically optimized. The obtained results for both length (L_{R1} and L_{R2}) are depicted in figure (4.25). It is observed in figure (4.25 a) that with increasing first stage length L_{R1} the resolution (at the detector plane) of the configuration decreases linearly (indicating the benefit of a strong deceleration of the ions before reflection). Contrary to this the stage length L_{R2} has no influence on resolution (for a fixed energy spread of the beam with $T_\perp = 1$ K). For a fixed first stage length ($L_{R1} = 12$ mm) the variation of the second stage length L_{R2} results in the same resolution value ($R = 2.59 \times 10^6$) whereas only the optimized potential for the second stage (U_{R2}) increases with the length L_{R2} linearly (see figure 4.25 b). However, it must be kept in mind that the high voltage power supplies in use deliver a maximum voltage value of 6 kV. Hence the first acceleration stage length L_{R1} can be designed as short as possible. Here the limit is given by technical problems related with short electrode distances (sparkovers and mesh flexing). To be on the save side the length L_{R1} was set to $L_{R1} = 12$ mm as used for the calculations depicted in figure (4.25 b). Additionally in figure (4.25 b) it can be observed that with decreasing second stage length L_{R2} the optimal voltage value decreases, too. For too short second stage length L_{R2} e.g. below $L_{R2} = 70$ mm the optimal voltage value of U_{R2} drops below 6 kV which would be the highest available kinetic energy of the accelerated ions. In that case these ions would not be reflected in the second reflectron stage. Hence $L_{R2} = 70$ mm would be the lower limit of the second reflectron stage length L_{R2}. Keeping this restriction in mind the second task was to find a suitable design for the "real" reflectron. In the section about the deflector (4.1.3) an optimized reflectron configuration was used. Here we will describe the optimization process of this reflectron which was implemented in the experimental setup later (see figure 3.1). The first step was to check the integrity of the results obtained by numerical optimization with SIMION simulations. Therefore the ideal accelerator configuration (similar as in figure 4.7) and an ideal reflectron configuration consisting of three grids were used. The SIMION simulations delivered the same results as obtained by numerical optimization. The ions were flown without a transversal velocity component (one dimensional motion). Hence in SIMION a user program was written which turned off the acceleration potentials after ion extraction. In that case reflected ions flew through the accelerator to the detector which was positioned behind the accelerator. After the confirmation of the numerical optimization results with SIMION simulations we proceeded with the optimization of

4.1 TOFMS Optimization

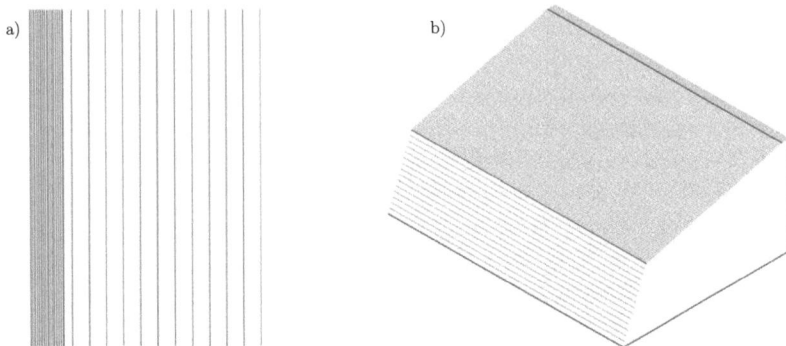

Figure 4.26 SIMION simulation of the "ideal" two stage reflectron consisting of three meshes ($L_{R1} = 12$ mm, $L_{R2} = 70$ mm, $\varnothing = 120$ mm). Optimized geometries of the accelerator and deflector were used with numerically optimized potential values. The obtained resolution for Ar_{25}^+ ions with accelerator spatial focus plane at the mass gate $L_F = 232.6$ mm is $R = 16221$ for 5 kV ion extraction (R at the detector). **a)** SIMION electrode view with equipotential lines (red) of the two stage ideal reflectron. Here the meshes are drawn up to the boundaries of the potential array whereas SIMION assumes that the meshes are extended to infinity. Hence the equipotential lines in the reflectron stages are parallel indicating ideal field homogeneity. **b)** Potential energy surface view of the two stage "ideal" reflectron shown in a).

the reflectron. Therefore similar simulations as described in the section about the deflector were used to find the optimal reflectron configuration. Analogously to the deflector SIMION simulations an Argon (Ar_{25}^+) ion matrix (2 × 20) mm was started and the TOF distribution was used for the calculation of the resolution (calculated by the time dispersion at the detector see equation 2.37). For the SIMION simulations the optimized three stage accelerator geometry (see figure 4.9) and the optimized deflector geometry (see figure 4.19) with optimal voltages were used. The deflector potential (bipolar) was adjusted for deflecting all ions to the detector where the TOF of the ions is recorded. SIMION simulations of the "ideal" reflectron consisting of three meshes delivered the maximum (reference) resolution for the Re-TOFMS setup. The configuration of this "ideal" reflectron geometry is depicted in figure (4.26). With the "ideal" reflectron configuration a resolution of $R = 16221$ was obtained by SIMION simulations with the Ar_{25}^+ ion matrix at 5 kV extraction. Here as in the case of the "ideal" accelerator (see figure 4.7) SIMION assumes that the meshes are extended to infinity (no boundary effects). With the use of an outer shielding and electrodes of "real" dimensions the situation changes dramatically as in the case of the accelerator (see section 4.1.2).

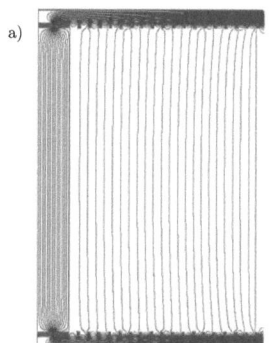
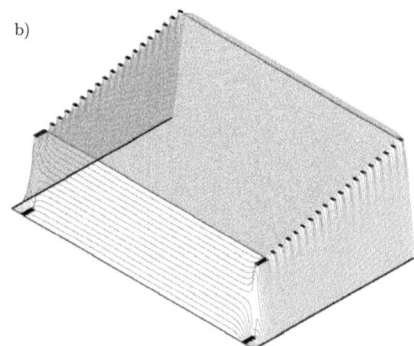

Figure 4.27 SIMION simulation of a "real" two stage reflectron consisting of two stages (deceleration stage: $L_{R1} = 12$ mm, reflection stage: $L_{R2} = 71$ mm, $\varnothing = 120$ mm). Optimized geometries of the accelerator and deflector were used with numerically optimized potential values. The obtained resolution for Ar_{25}^+ ions with accelerator spatial focus plane at the mass gate $L_F = 232.6$ mm is $R = 2012$ for 5 kV ion extraction (R at the detector). **a)** SIMION electrode view with equipotential lines (red) of the two stage reflectron. Here the equipotential lines in the reflectron stages are not parallel as in the case shown in figure 4.26 indicating electric potential distortions. **b)** Potential energy surface view of the reflectron configuration shown in a).

Within the optimization process of the accelerator geometry (see section 4.1.2) it was shown that "pot" shaped electrodes are well suited for short stage length. Hence the first deceleration stage of the reflectron can consist of a "pot" shape electrode ($L_{R1} = 12$ mm). Here the question arises for the configuration of the second reflectron stage which is much longer than the first deceleration stage. The first attempt was to use many ring electrodes for the second reflection stage (as in the case of the third acceleration stage see section 4.1.2). One possible two stage reflectron configuration with a "pot" shaped first deceleration stage and a second stage with ring electrodes is depicted in figure (4.27). For the second stage of the configuration depicted in figure (4.27) 17 ring electrodes were used (with 1 mm thickness and 18×3 mm spacing between the electrodes). The "pot" shaped configuration of the first stage delivers a homogeneous field distribution whereas the equipotential lines in the second stage are not perfectly parallel indicating field inhomogeneity (see the curvature of the equipotential field lines near the electrodes and the rear part of the reflectron in figure 4.27 a). Hence the obtained resolution of $R = 2012$ is far away from the ideal value of $R = 16221$ (see figure 4.26). This can be explained by the "low" number of ring electrodes used in figure (4.27). Therefore in the next configuration the number of the ring electrodes was

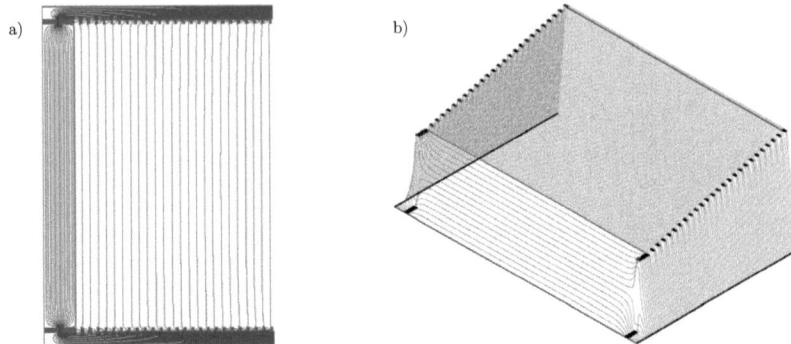

Figure 4.28 SIMION simulation of a "real" two stage reflectron consisting of two stages (deceleration stage: $L_{R1} = 12$ mm, reflection stage: $L_{R2} = 74$ mm, $\varnothing = 120$ mm). Optimized geometries of the accelerator and deflector were used with numerically optimized potential values. The obtained resolution for Ar_{25}^+ ions with accelerator spatial focus plane at the mass gate $L_F = 232.6$ mm is $R = 8166$ for 5 kV ion extraction (R at the detector). **a)** SIMION electrode view with equipotential lines (red) of the two stage reflectron. Here the equipotential lines in the reflectron stages are nearly parallel indicating better field homogeneity than in figure 4.27. **b)** Potential energy surface view of the same reflectron configuration shown in a).

increased from 20 to 27 which additionally reduce the spacing between the electrodes to 2 mm. The resulting SIMION configuration with equipotential lines in depicted in figure (4.28). The second stage of the configuration depicted in figure (4.28) consists of 27 ring electrodes (with 1 mm thickness and 2 mm spacing between the electrodes). In this case the resolution of the whole Re-TOFMS setup increases about a factor of 4 to $R = 8166$ which is still half of the "ideal" configuration depicted in figure (4.26). Here the question arises if a better resolution can be obtained by a similar electrode configuration. Therefore it was decided to simulate configurations which obey a certain symmetry e.g. same value for electrode thickness and spacing between the electrodes. Such a configuration with electrode thickness equal to electrode spacing (23 electrodes with 1.6 mm thickness and 24×1.6 mm spacing between the electrodes) is depicted in figure (4.29). This configuration delivers the same resolution value as the "ideal" reflectron (4.26). With the configuration in figure (4.29) the optimization process could be finished. This result can be attributed to the symmetric configuration with equal value for electrode thickness and spacing between the electrodes. This result was verified with other configurations which fulfilled the same symmetry requirement. By a reflectron configuration ($L_{R1} = 12$ mm, $L_{R2} = 72$ mm, $\varnothing = 120$ mm and 20 electrodes) with 2 mm electrode thickness and 2 mm spacing between the electrodes

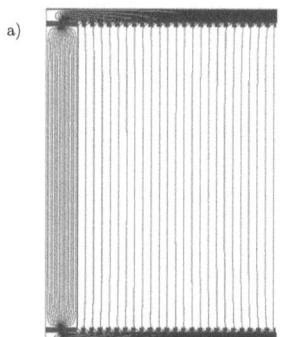 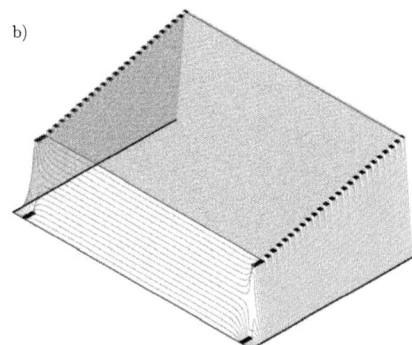

Figure 4.29 SIMION simulation of a "real" two stage reflectron consisting of two stages (deceleration stage: $L_{R1} = 12$ mm, reflection stage: $L_{R2} = 75.2$ mm, $\varnothing = 120$ mm). Optimized geometries of the accelerator and deflector were used with numerically optimized potential values. The obtained resolution for Ar_{25}^+ ions with accelerator spatial focus plane at the mass gate $L_F = 232.6$ mm is $R = 16220$ for 5 kV ion extraction (R at the detector). **a)** SIMION electrode view with equipotential lines (red) of the two stage reflectron. Here the equipotential lines in the reflectron stages are nearly ideally parallel indicating nearly ideal field homogeneity as shown in figure 4.26. **b)** Potential energy surface view of the same reflectron configuration shown in a).

a resolution of $R = 9034$ was obtained which is better than the configuration with 27 electrodes (see figure 4.28). However, in the case when electrode thickness and spacing between the electrodes exceeded 2 mm requiring at least 20 electrodes the resolution of the setup decreased drastically. However it was interesting to search for configurations with equal performance and less electrodes than the 26 electrodes used for the configuration in figure (4.29). Therefore other electrode configurations were simulated. Other interesting configurations which were simulated were configurations with alternating thicker electrodes followed by thinner electrodes. The resulting field distribution for such electrode configurations are comparable with "pot" shaped electrodes. In this case the number of electrodes can be reduced without decreasing resolution dramatically. Such an alternating electrode configuration is depicted in figure (4.30). The obtained resolution with the configuration depicted in figure (4.30) is higher than the resolution obtained with 27 separate electrodes (see figure 4.28) whereas only 6 separate electrodes are in use. The obtained equipotential field distribution is comparable with the field distribution of two "pot" shaped electrodes whereas no additional meshes are implemented. By further optimization of this alternating electrode configuration the resolution of the "ideal" reflectron configuration with much less electrodes

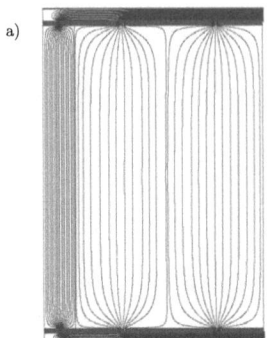 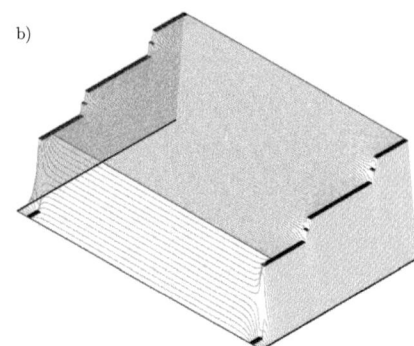

Figure 4.30 SIMION simulation of a "real" two stage reflectron consisting of two stages (deceleration stage: $L_{R1} = 12$ mm, reflection stage: $L_{R2} = 72$ mm, $\varnothing = 120$ mm) with alternating electrode thickness geometry. The first and last electrode have the half thickness (16.5 mm) of the middle electrode (33 mm). The thinner electrodes are 1 mm thick. Between all electrodes the same spacing of 1 mm is used. Optimized geometries of the accelerator and deflector were used with numerically optimized potential values. The obtained resolution for Ar_{25}^+ ions with accelerator spatial focus plane at the mass gate $L_F = 232.6$ mm is $R = 9301$ for 5 kV ion extraction (R at the detector). **a)** SIMION electrode view with equipotential lines (red) of the two stage reflectron. Similar field distribution as in the case of "pot" shaped electrodes. **b)** Potential energy surface view of the same reflectron configuration shown in a).

(12 electrodes) was realized. It must be emphasized that these configurations must obey a certain symmetry requirement. The first and the last electrode of the stage must have the half length of the thicker electrodes (e.g. 6.5 mm in the case of 13 mm thick electrodes). Best results were obtained where the spacing between the electrodes was equal to the thickness of the thinner electrode (e.g. 0.5 mm spacing between the electrodes for 0.5 mm thick thinner electrodes). The final SIMION simulation of the fully optimized two stage reflectron is depicted in figure (4.31). This configuration was used in the simulations for the optimization of the deflector geometry (see section 4.1.3). It must be mentioned here that to our knowledge up today similar configurations are not reported in literature. With such a reflectron configuration the number of electrodes can be reduced significantly (simplifying construction) without suffering a loss in resolution. A picture of the constructed reflectron and other components can be found in the appendix A. Here figure 4.32 shows a section view of the final reflectron collider configuration based on the reflectron geometry as shown in figure 4.31. In figure 4.32 the whole reflectron assembly is mounted with M8 stainless steel rods on a DN200CF flange (zero length reducer to DN100CF). The reflectron collider was

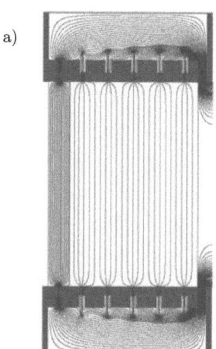 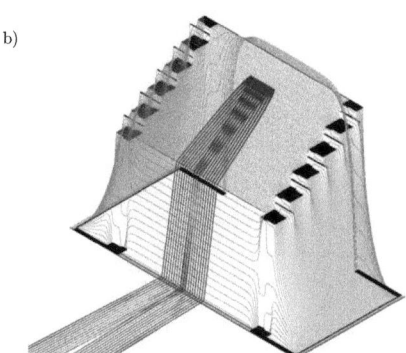

Figure 4.31 SIMION simulation of a "real" two stage reflectron consisting of two stages (deceleration stage: $L_{R1} = 12$ mm, reflection stage: $L_{R2} = 72.5$ mm, $\varnothing = 116$ mm)with alternating electrode thickness geometry. The first and last electrode have the half thickness (6.5 mm) of the middle electrodes (13 mm). The thinner electrodes are 0.5 mm thick. Between all electrodes the same spacing of 0.5 mm is used. Optimized geometries of the accelerator and deflector were used with numerically optimized potential values. The obtained resolution for Ar_{25}^+ ions with accelerator spatial focus plane at the mass gate $L_F = 232.6$ mm is $R = 16221$ for 5 kV ion extraction (R at the detector). **a)** SIMION electrode view with equipotential lines (red) of the two stage reflectron. The field distribution in the second stage looks like as if five "pot" shape electrodes were arranged one after the other. The equipotential lines in the reflection stages are in the middle of the stage ideally parallel indicating high field homogeneity. **b)** Potential energy surface view of the reflectron configuration shown in a) with simulated ion trajectories (Ar_{25}^+) of the reflected ion matrix (2×20 mm).

utilized in two different surface impact configurations. In the prior configuration the cluster ions were impacted on the stainless steel backplane of the reflectron collider. Therefore the last electrode mesh which defines the potential U_{R2} was replaced by the scattering surface consisting of a polished circular 0.5 mm stainless steel plate. This configuration was further improved by the design of a new surface holder which can hold different surfaces (e. g. silicon, see figure 4.32). The surface holder is a circular stainless steel disc with 0.5 mm thickness and a rectangular hole (100 mm × 20 mm) in the middle for the surface sample. Fine notches on the rectangular hole edges of the surface holder (with 0.1 mm thickness for low electric field distortion inside the reflectron) and spring plate clamps (behind the surface) keep the silicon surface in its place. This surface holder is placed in a distance of 0.8 mm behind the last electrode and mesh (as shown in figure 4.32). Contrary to the stainless steel plate due to insulation with alumina ceramic spacers from the rest of the reflectron the silicon surface and surface holder can be set to a different potential U_S. Further for the desorption of water molecules additionally a surface heater (GY 6.35 base halogen filament lamp with maximal 150 W heating power) was implemented behind the surface. The surface heater is mounted on a DN100CF flange (zero length reducer to DN40CF) allowing easy lamp change. Silicon surface samples were laser cut to rectangular shape plates (100 mm × 20 mm) from a ⌀ = 150 mm silicon wafer disc[2]. The silicon wafer disc was a p-type (boron doped) disc with $675 \pm 25 \mu$m thickness. One surface side is "perfectly" polished whereas the other side is chemically etched and matt. The polished surface side with (100)-orientation was used for the cluster surface impact experiments. Resistance of the silicon surface sample is with $1 - 30$ Ωcm high enough to avoid charging of the surface by the ion beam. Due to the higher reactivity of pure silicon the surface is left with its natural oxide SiO_2 passivation layer (several nanometers of thickness).

---

[2]Donated by the Institut für Kristallzuechtung (IKZ), Dr. Helge Riemann.

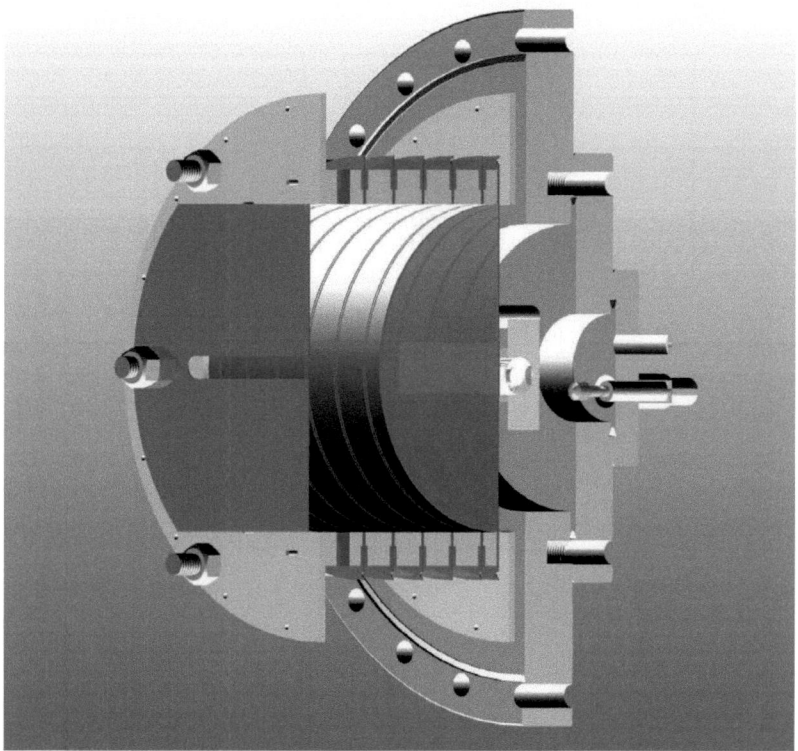

Figure 4.32 Depicted is a section view of the reflectron collider with the reflectron support mounted on a DN200CF flange (zero length reducer to DN100CF). The configuration of the reflectron collider is based on the optimized geometry shown in figure 4.31. Depicted is the configuration with a silicon surface placed in the surface holder (the last electrode hold at U_S potential). The surface holder is a circular stainless steel disc (0.5 mm thickness) with a rectangular hole for the surface (100 mm × 20 mm). Fine notches on the edges of the surface holder (with 0.1 mm thickness for low electric field distortion inside the reflectron) and spring plate clamps keep the silicon surface in its place. For surface heating a tungsten filament halogen lamp is placed behind the surface. Additional pictures of the reflectron are shown in the appendix A (figure A.4).

4.2 Time-of-Flight Mass Spectra

4.2.1 Mass Calibration

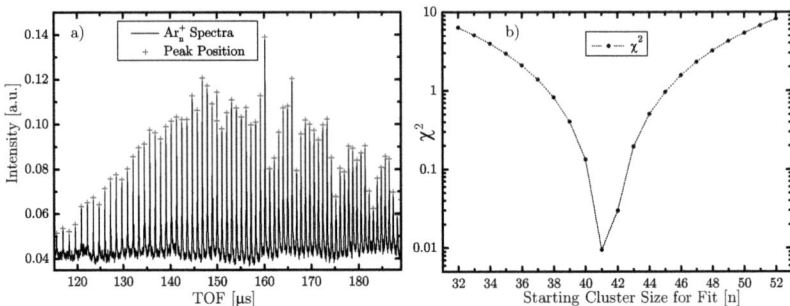

Figure 4.33 a) Depicted is the TOF mass spectra of neat Argon for 3 kV acceleration ($T_0 = 283$ K and $P_0 = 3$ MPa). The positions of the cluster size peaks were determined with the peak finding tool implemented in the software OriginPro ver. 8. These peak positions (red +) were used in the figure 4.34 for the mass calibration fits. **b)** By setting the parameter b in equation (4.1) equal to zero the beginning cluster size of the TOF spectra can be determined. In the case when b is set equal to zero the calibration fit (of the TOF peak position versus cluster size plot as in figure 4.34 a) delivers the lowest fit error χ^2 (•) for the beginning cluster size. In this case the beginning cluster size of the TOF spectra shown in a is $n = 41$. Note the logarithmic scale of the χ^2 axis.

The data collected in a TOFMS experiment are the ion flight times to the MCP-detector. To determine the m/z ratio of the extracted ions a conversion from time domain to mass to charge (m/z) domain is required. The relation between the final TOF and the mass to charge ratio was described earlier in the subsection about numerical optimization (2.2.5). Using the equations from the subsection (2.2.5) the TOF can be converted to m/z values. However, the equations of subsection (2.2.5) give long and complicated terms for the calculation of the m/z values. Therefore alternatively the more basic TOF equation,

$$m/z = a(t_{\text{peak}})^2 + b \quad (4.1)$$

is often used for curve fitting in TOF mass spectrometry (where a and b are constants based upon instrumental parameters and t_{peak} the measured TOF for the different cluster sizes) [277]. For the calibration of cluster mass spectra "clean" TOF spectra e.g. of Ar where the presence of fragments do not complicate the spectra are preferred (see figure 4.33 a). In the first step the TOF of each cluster peak is plotted versus an estimated m/z value (cluster size see figure 4.34). In

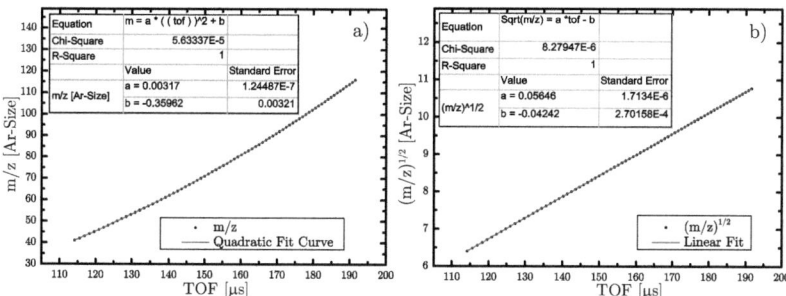

Figure 4.34 Mass calibration fits for conversion of TOF spectra to mass spectra. Ar cluster sizes plotted versus TOF peak positions (from figure 4.33) and fitted by two equivalent calibration functions. The TOF spectra was recorded with two stage 3 kV acceleration ($U_0 = 3$ kV, $U_1 = 2435$ V, $U_{R1} = 1788$ V and $U_{R2} = 2832$ V) for Ar gas expansion ($P_0 = 3$ MPa and $T_0 = 283$ K). **a)** Quadratic fit with equation (4.1) of Ar cluster size versus TOF plot points. The obtained calibration curve can be used for the conversion of TOF to m/z for the same TOFMS settings. **b)** An alternative calibration method for the conversion of TOF to m/z. Here the square root of the Ar cluster size is plotted versus the TOF. The resulting calibration curve is a linear function.

that case the equation (4.1) is used to fit the plotted points. Therefore the value b is initially set to zero and the cluster size is shifted stepwise ($n + 1$) to find the beginning cluster size (n). This fitting procedure is shown in figure 4.33 b) for the determination of the beginning cluster size $n = 41$ which delivers the lowest fit error χ^2. The resulting best fit (lowest χ^2 value) is then fitted again with the equation (4.1) using b also as an parameter (see figure 4.34). In the case when b is not set to zero the fit function is flexible enough to fit wrong beginning cluster sizes. Therefore for finding the beginning cluster size of the TOF spectra it is necessary to initially set $b = 0$. The obtained fit function is than the calibration function for the TOF spectra with identical settings (accelerator and reflectron voltages). Alternatively the TOF can be plotted versus the square root of the estimated m/z values. In this case the calibration curve would be linear. Both equivalent calibration methods are depicted in the figure (4.34 a and b). For the calibration fits an Ar TOF spectra was used. The TOF data in figure (4.34 a and b) is well fitted with both functions over a large mass range starting with Ar_{40}^+ (1597.92 amu) up to Ar_{120}^+ (4793.76 amu). In the case of $(CO)_n^+$ and $(CO_2)_n^+$ attention should be paid to the isotopic distribution (see also 4.2.3). For bigger clusters the probability to contain ^{13}C isotopes or ^{18}O isotopes growths with the size of the cluster. In that sense for e.g. $(CO_2)_n^+$ clusters bigger than $N = 86$ and $(CO)_n^+$ clusters bigger than $N = 89$ the isotope peak with $+ 1$ amu is more

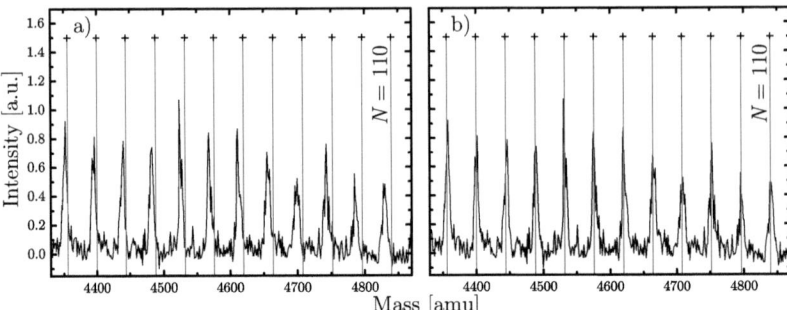

Figure 4.35 The same TOF spectra calibrated with two different mass calibration fits. $(CO_2)_n^+$ cluster sizes ranging from $N = 99$ up to $N = 110$ plotted versus the calibrated mass in amu. The corresponding TOF spectra was recorded with two stage 2950 V acceleration in linear TOFMS configuration ($U_0 = 2950$ kV, $U_1 = 2665$ V) for CO_2 gas expansion ($P_0 = 7.5$ MPa seeded in Ar (1:5 ratio) and $T_0 = 298$ K). The mass spectra in **a)** was calibrated with a fit function obtained for the whole mass range ($N = 5$ up to $N = 120$, without correction for the higher intensity of the $N + 1$ amu isotopic peak. The crosses with vertical lines (red) show the expected exact mass positions for the $N + 1$ amu isotopic peak. The discrepancies between the expected mass peak positions and the calibrated mass peaks increase with N up to 12 amu for $N = 110$. **b)** Shows the same mass spectra calibrated with a fit function obtained for the high mass range ($N = 87$ up to $N = 120$ corrected for the higher intensity of the $N + 1$ amu isotopic peak). The crosses with vertical lines (red) show the expected exact mass positions for the $N + 1$ amu isotopic peak. No discrepancies between the expected mass peak positions and the calibrated mass peaks can be observed.

intense than the usual ^{12}C peak. For these cluster sizes the peak maximum which is determined with peak finding software or tools is than shifted from n to $n + 1$ amu. Therefore two different mass calibration curves, one for the low mass region with n and one for the high mass region with $n + 1$ amu will be necessary (for $N \leq 86$ and $N \geq 87$ for $(CO_2)_n^+$ clusters or $N \leq 89$ and $N \geq 90$ for $(CO)_n^+$ clusters). Otherwise discrepancies between the expected exact mass and the calibrated mass spectra arise for large clusters which is e. g. shown in figure 4.35 a) for $(CO_2)_n^+$ clusters with $N = 99 - 110$. These discrepancies can be eliminated by the use of a second mass calibration and taking into account the higher isotope intensity of the high mass range as shown for the case of $(CO_2)_n^+$ clusters for $N \geq 87$ (see figure 4.35 b). The isotope distribution depends on the relative abundance of the isotopes (e. g. $p = 1.1\%$ for ^{13}C and $p = 0.2\%$ for ^{18}O) and can be calculated by the binomial distribution of the isotopes. In that sense the isotopic distribution for C_n can be calculated by the equation

$$B(k|p,n) = \binom{n}{k} p^k (1-p)^{n-k}, \qquad (4.2)$$

where k defines the number of isotopes in a cluster with n carbon atoms. Additionally for the calculation of complicated isotopic distributions computer programs are available (implemented in most molecular weight calculation programs) which are also distributed as freeware[3]. In high resolution mass spectra the isotopic distribution can be resolved which will be discussed later on in the subsection (4.2.3).

4.2.2 Linear TOFMS Mass Resolution

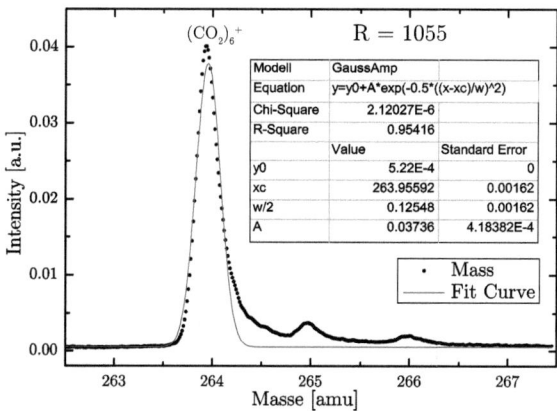

Figure 4.36 Mass resolution of the two stage linear TOFMS configuration operated at 3 kV extraction (U_0 = 2950 V, U_1 = 2679 V and 70 eV EI). Zoomed view of the most intense peak $(CO_2)_6^+$ in the mass spectra of $(CO_2)_n^+$ cluster ions. The mass resolution was determined by fitting the main $(CO_2)_6^+$ mass peak with an Gaussian fit curve (red curve) to determine the FWHM of the peak. Due to the high resolution of the configuration (R = 1055) the less intense isotope peaks with +1 amu difference are also visible in the mass spectra. The (CO_2) was expanded seeded in Ar (1:5 ratio, P_0 = 8.4 MPa and T_0 = 298 K)

Before the reflectron was implemented into the setup the proper function of the TOFMS accelerator was checked. Therefore the TOFMS was operated in linear configuration with the MCP detector positioned in place of the reflectron (see figure 3.1). The optimal extraction delay for the accelerator was determined by

---

[3] e. g. Molecular Weight Calculator by Matthew Monroe, www.alchemistmatt.com

observing the molecular beam intensity with the Faraday cup. After some test runs of the accelerator and deflector the resolution of the system was optimized by adjusting acceleration and deflection voltages (space focus at the detector plane with $L_D \approx 712$ mm). Due to the higher cluster ion intensities of $(CO_2)_n^+$ clusters, CO_2 seeded in Ar (1:5 ratio) was used as the expansion gas instead of neat Ar. The most intense peak was chosen for determination of the mass resolution. The recorded mass peak for the most intense cluster size $(CO_2)_6^+$ is depicted in figure (4.36). The main mass peak of $(CO_2)_6^+$ was fitted by a Gaussian fit curve (red curve in figure 4.36) to determine the full width at half maximum (FWHM). From the full width at half maximum ($\Delta m_{FWHM} = 0.251$ amu) of the Gaussian peak and the center mass of $m = 263.96$ amu a mass resolution of $R = 1055$ was derived, despite the TOFMS was operated in linear two stage configuration at 3 kV extraction.

4.2.3 Reflectron TOFMS Mass Resolution

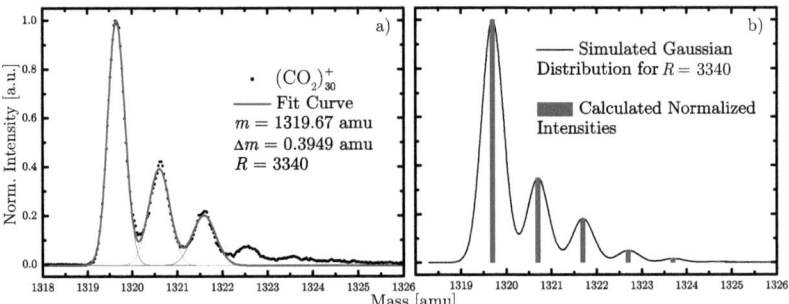

Figure 4.37 a) Mass resolution of the two stage reflectron TOFMS configuration operated at 4 kV extraction ($U_0 = 4$ kV, $U_1 = 3505$ V, $U_{R1} = 2508$ V, $U_{R2} = 3970$ V and 250 eV EI). Zoomed and normalized view of one of the most intense peaks ($(CO_2)_{30}^+$) in the mass spectra of $(CO_2)_n^+$ cluster ions. The mass resolution was determined by the FWHM of the first peak of the multiple Gaussian peak fits of the isotopic distribution (brown curve). Due to the cluster size and high resolution of the configuration ($R = 3340$) the less intense isotope peaks with up to +3 amu difference are also visible in the mass spectra. The sample gas (CO_2) was expanded seeded in Ar (1:5 ratio, $P_0 = 4.1$ MPa and $T_0 = 303$ K). b) Depicted is for comparison the isotopic mass distribution of $(CO_2)_{30}$ calculated with the molecular weight calculator (see 4.2.1). Shown are the calculated normalized intensities of the isotopes (bar graph) and the simulated Gaussian distribution for an estimated resolution $R = 3340$ (line graph).

After the successful operation of the linear TOFMS configuration the setup was expanded with the ion reflectron. Therefore the MCP detector was moved behind the accelerator and the reflectron was placed inside the 100 mm extension tube ($\varnothing \approx 200$ mm) as shown in figure (3.1). Here again the resolving power of the reflectron TOFMS configuration was tuned after a few test runs. High resolution mass spectra of $(CO_2)_n^+$ cluster ions were recorded for the determination of the resolving power of the apparatus in reflection mode. A representative result of such an isotopic resolved high resolution mass spectra of $(CO_2)_{30}^+$ cluster ions is depicted in figure (4.37). The sample gas (CO_2) was seeded in Ar (1 : 5 ratio) as in the case of the linear configuration (see 4.2.2). The reflectron TOFMS was operated in two stage configuration without deflection and mass selection at 4 kV acceleration. Hence, the space focus plane was shifted to the detection plane ($L_D \approx 1455$ mm). The resolution of the configuration was determined from the FWHM ($\Delta m = 0.3949$ amu) of the most intense isotope mass peak ($m = 1319.67$ amu) obtained by multiple Gaussian peak fit curves. In that sense the resolution of the apparatus exceeds $R = 3000$. Due to the high resolution the isotopic mass distribution of the $(CO_2)_{30}^+$ cluster ions are well resolved (isotopic masses up to +3 amu). Additionally the recorded isotopic distribution matches well with the calculated distribution shown in figure (4.37 b).

4.2.4 Mass Separation

The mass selection performance of the mass gate was tested in two different configurations. With and without shielding meshes mounted in front of the ion entrance and exit slits. In the case shielding meshes mounted in front of the slits the mass selection performance of the gate is much higher than without shielding meshes. With the configuration with shielding meshes cluster sizes around $N = 190$ could be mass selected in excess. A representative mass spectra of mass selected $(CO_2)_{190}^+$ recorded with this configuration is shown in figure (4.38). However, the cluster ion intensity is reduced due to the use of two additional meshes for the mass gate. Therefore the question arises for the mass selection performance of the mass gate for the configuration without shielding meshes. Hence, the meshes were removed from the mass gate to test the mass selection performance without meshes. In that case the mass selection performs is reduced but is in excess higher than $N = 120$ without reducing cluster ion intensity. These result encouraged us to use the mass gate without shielding meshes and suffering a loss in mass selection performance than reducing the ion intensity. However, in the case where higher mass selection performance is needed the shielding meshes can be mounted again on the mass gate.

4.2.5 Cluster Size and Intensity

The cluster size distribution in the molecular beam depends on many factors described before in the subsection (2.1.3). By the variation of these parameters

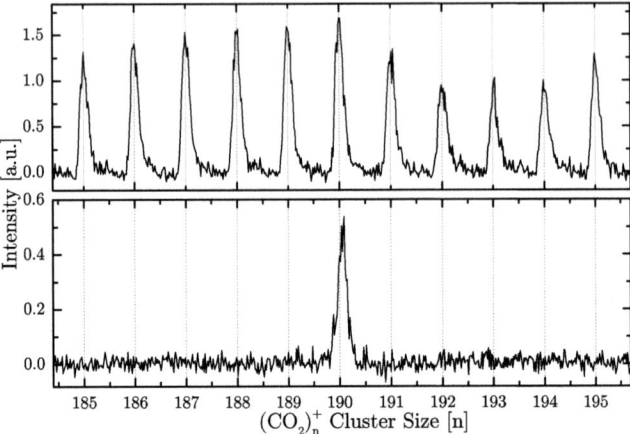

Figure 4.38 Mass selection of a big $(CO_2)_n^+$ cluster ion with $N = 190$. The space focus is located at the mass gate wire plane. The reflectron TOFMS is operated in two stage configuration at 4 kV extraction ($U_0 = 4$ kV, $U_1 = 3248$ V, $U_{R1} = 2562$ V, $U_{R2} = 4082$ V and 200 eV EI). Shielding meshes were mounted in front of the mass gate entrance and exit slits. The upper mass spectra shows the zoomed view of the unfiltered mass spectra of $(CO_2)_n^+$ with $N = 185 - 195$ (mass gate off). The lower mass spectra shows the filtered mass peak of $(CO_2)_{190}^+$ cluster ions (mass gate on, ±220V pulsed). The sample gas (CO_2) was expanded seeded in Ar (1:5 ratio, $P_0 = 5$ MPa and $T_0 = 298$ K).

the cluster size distribution can be shifted to bigger or smaller clusters. However, during the measurement it is not possible to change some of these parameters (e.g. nozzle diameter or expansion half angle) or it is not wanted to change some of these parameters (e. g. due to sluggish temperature changes). Hence other parameters were chosen to influence the cluster size distribution with the aim to maximize the intensity of a desired cluster size. These cluster ions can be mass selected and used for further scattering experiments. In this subsection we will discuss the influence of these different parameters which affect the cluster size distribution and will show some exemplary results for the two different sample molecules CO and CO_2. Additionally with these results the need for the utilization of two different electron gun systems can be justified to generate different cluster size distributions. In that sense a valve mounted e-gun (3 in figure 3.1) with a fixed and lowest distance to the nozzle was used to generate cluster size distributions with big cluster ions. Contrary to this a flange mounted e-gun (2 in figure 3.1) with a variable distance to the nozzle which can be changed by the valve position was used to generate small clusters ions (beginning with the monomer).

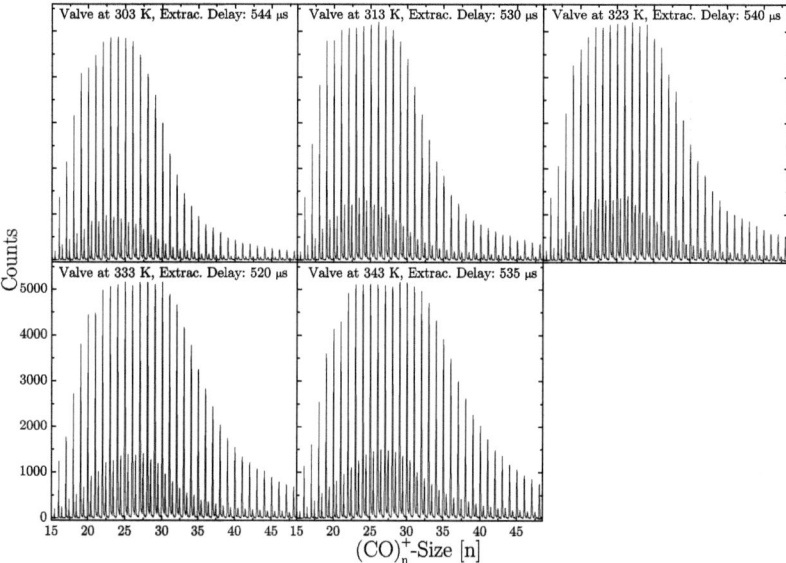

Figure 4.39 Depicted is the bimodal character of the cluster size distribution of $(CO)_{16}^+ - (CO)_{48}^+$ cluster ions in dependence of the valve temperature (all graphs with the same scale). Molecular beam expansion of neat CO. Valve was held at 2.5 MPa stagnation pressure and 7 Hz repetition. The valve mounted e-gun was used for ionization at 250 eV electron energy. Mass spectra recorded at 3 kV acceleration and 2.5 µs extraction pulse for the Re-TOFMS. The extraction delay between valve opening and TOFMS extraction was optimized for maximal intensity for respective temperatures (520µs–544µs). Deflection plates were used bipolar with ±10 V deflection for increasing the intensity of small clusters.

Stagnation temperature The stagnation temperature is one of the values which will remain constant during the measurement and is not wanted to be changed. However, it is interesting to investigate its influence on the cluster size distribution. Therefore the stagnation temperature of the sample gas was varied and the resulting cluster size distribution was observed. The development of the cluster size distribution in dependence of the stagnation temperature is depicted in figure (4.39). The valve temperature was increased up to 343 K. Hence the extraction delay time of the TOFMS was optimized to follow the increasing mean beam velocity (see equation (2.5)). With increasing valve temperature also increasing fragment intensity can be observed. Bigger size clusters gain intensity due to the fact that the size distribution becomes broader. For example the cluster $(CO)_{35}^+$

of medium size gains a factor of four on count rate. These effects can be attributed to the temperature dependent opening characteristics of the valve and temperature dependent particle flow.

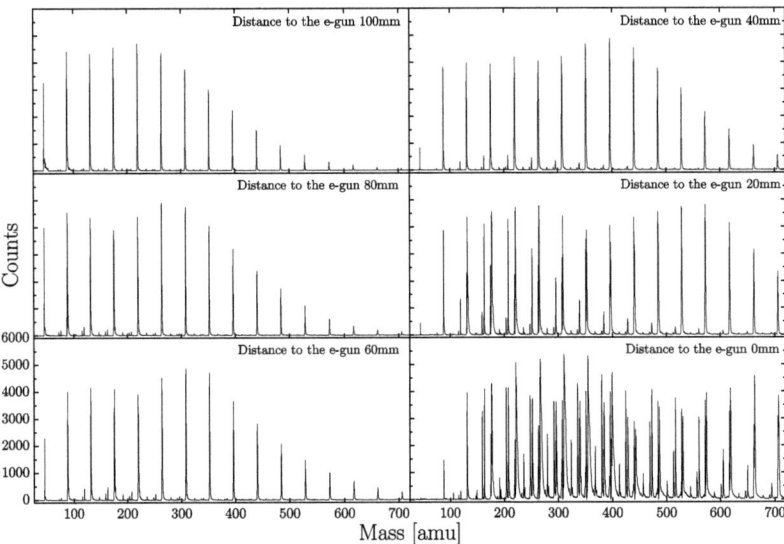

Figure 4.40 Mass spectra of $(CO_2)^+ - (CO_2)_{16}^+$ clusters in dependence of valve to e-gun distance (all graph with the same scale). Seeded beam of CO_2 in He with a (1 : 2) ratio. Valve was held at 304 K and 3.5 MPa stagnation pressure and 7 Hz repetition. The flange mounted e-gun was used for ionization at 300 eV electron energy. Mass spectra recorded at 3 kV acceleration and 2.5 μs extraction pulse for the Re-TOFMS. The extraction delay between valve opening and TOFMS extraction was optimized for maximal intensity for respective distances (510 μs - 610 μs). Deflection plates were used bipolar with ±20 V deflection for increasing the intensity of small clusters.

Distance between the e-gun and nozzle Another parameter which influences the cluster size distribution is the distance between the cluster source (nozzle) and the electron source (see subsection 3.2.2) for ionization. The relative distance between the nozzle exit and the orifice of the shielding electrode of the e-gun was determined with the valve manipulator scale (approximately ±1 mm accuracy). The intensity of a desired cluster size can be well controlled by this parameter. The change in cluster size distribution in dependence of the e-gun to nozzle distance for CO_2 is depicted in figure (4.40). The distance between

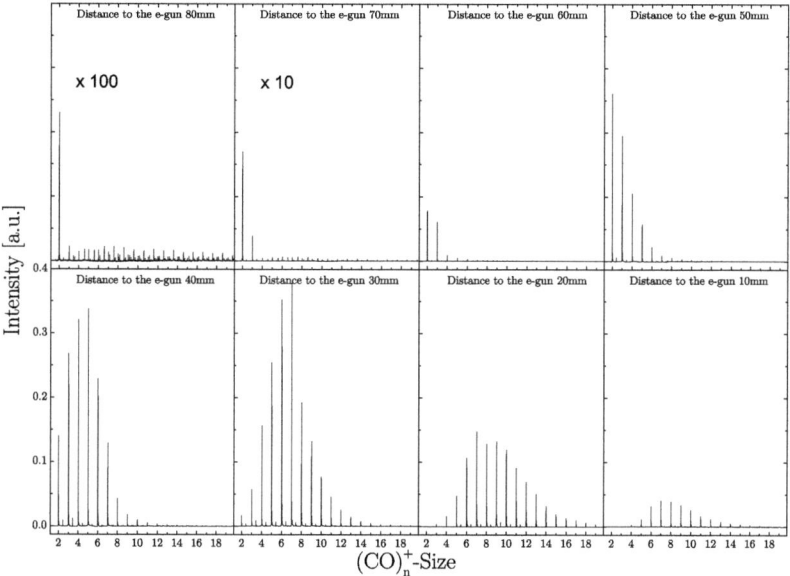

Figure 4.41 Mass spectra of $(CO)_2^+ - (CO)_{19}^+$ clusters in dependence of valve to e-gun distance. Molecular beam expansion of neat CO. Valve was held at 303 K and 4 MPa stagnation pressure. The flange mounted e-gun was used for ionization at 70 eV electron energy. Mass spectra were recorded at 3 kV acceleration and 15 µs extraction pulse for the reflectron TOFMS. The extraction delay between valve opening and TOFMS extraction was optimized for maximal intensity for respective distances (635 µs – 710 µs). Deflection plates were used bipolar with ±35 V deflection for increasing the intensity of small clusters.

the valve and the e-gun was gradually reduced from 100 mm to zero distance. At larger distances clean mass spectra were observed. Contrary decreasing the distance between the valve and e-gun resulted in increasing fragment ion yields. Additionally the mass resolution is decreased by peak broadening and tailing of the peaks indicating fragmentation in the acceleration region of the TOFMS (TOF of the ion package to the TOFMS accelerator is reduced with the distance). Johnson et al. described the peak broadening by collision-induced dissociation of the higher clusters, yielding fragment ions [264]. According to their mass the fragments could be identified as $(CO_2)_n^+C$, $(CO_2)_n^+O$, $(CO_2)_n^+CO$ and $(CO_2)_n^+O_2$ (see also [106; 163; 278; 279]). The sharp increase in fragment intensities at low distances could be an effect of molecule ion reactions [280] caused by the higher beam flux density in front of the valve. In the case of CO the influence of the distance between the nozzle and the e-gun on the cluster size distribution is more pronounced than in the case of CO_2. The influence of the distance between the nozzle and e-gun on the cluster size distribution is depicted in figure (4.41). In figure (4.41) the distance between the valve and the e-gun was gradually reduced from 80 mm to 10 mm. At larger distances the dimer and small CO cluster ions dominate the mass spectra. Decreasing the distance between the valve and e-gun increases the intensities of larger clusters. Note the low fragment intensities in the mass spectra for the low electron energy applied. This effect can be explained by that fact that ionized species can act as condensation cores for the buildup of bigger clusters [264]. Thus bigger clusters seem to be more stable when formed at lower distances to the electron source. As opposed to this with increasing distance bigger clusters can be fragmented by electron bombardment increasing the amount of smaller cluster ions.

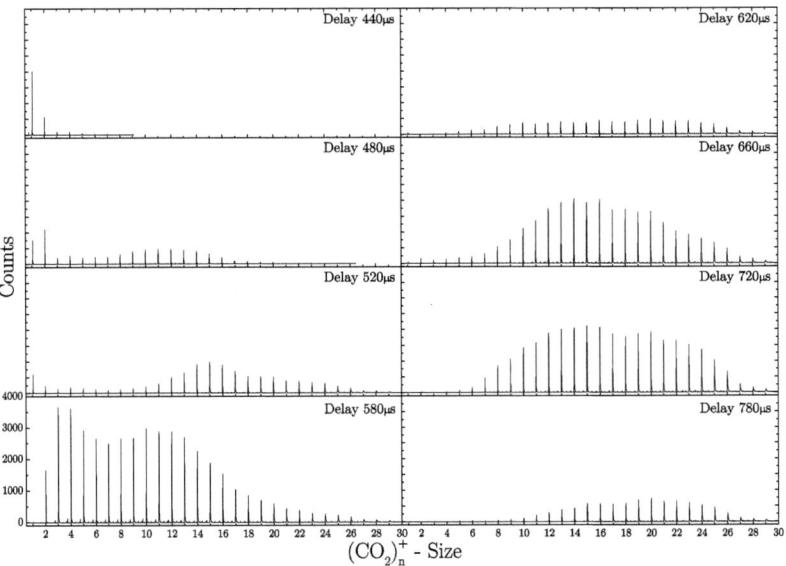

Figure 4.42 Mass spectra of $(CO_2)^+ - (CO_2)_{30}^+$ clusters in dependence of the extraction delay between valve opening and TOFMS acceleration voltage pulse (2.5 μs). Valve to e-gun distance fixed at 20 mm. Seeded beam of CO_2 in He with a 1 : 2 ratio. Valve was held at 304 K and 3 MPa stagnation pressure and 7 Hz repetition new e-gun was used for ionization at 300 eV electron energy. Mass spectra recorded at 3 kV acceleration and 2.5 μs extraction pulse for the reflectron TOFMS. Deflection plates were not used and were grounded to zero potential.

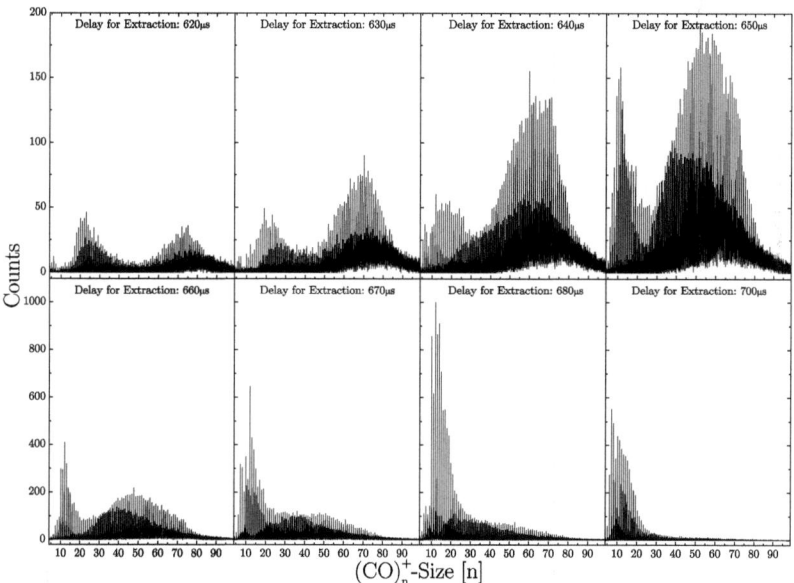

Figure 4.43 Mass spectra of $(CO)_4^+ - (CO)_{99}^+$ clusters in dependence of the extraction delay between valve opening and TOFMS acceleration voltage pulse. Molecular beam expansion of neat CO. Valve was held at 305 K and 2.6 MPa stagnation pressure. The flange mounted e-gun was used for ionization at 250 eV electron energy. Mass spectra recorded at 4 kV acceleration and 2.5 μs extraction pulse for the reflectron TOFMS. Deflection plates were grounded and not used.

Time delay between valve opening and extraction In pulsed nozzle operation the cluster size distribution varies along the beam pulse [90]. Thus by variation of the extraction delay different portions of the beam pulse with different cluster size distributions can be extracted. Therefore the delay between the valve opening and the extraction delay of the accelerator was gradually scanned from 440 μs up to 780 μs for the CO_2 sample gas (see figure 4.42). At early delay times more small clusters can be located in the mass spectra. The mean intensity maximum shifts from small clusters to bigger clusters. At a delay time of 580 μs the intensity maximum for medium size clusters ($N = 12$) is reached. Increasing the delay time further leads to a lower intensity for all cluster sizes followed afterwards by an increase of the intensity of bigger sized clusters. This behavior is much more pronounced in the mass spectra for CO (due to the higher extraction potential of 4 kV, see figure 4.43). In the case of CO the delay between the valve opening and the extraction delay of the accelerator was gradually scanned from

620 μs up to 700 μs. At early delay times a bimodal cluster size distribution with maxima around $(CO)_{23}^+$ and $(CO)_{70}^+$ can be seen. With increasing delay times both distributions gain intensity and shift to smaller clusters. At a delay time of 660 μs the intensity maximum for the medium size clusters ($N \approx 50$) is reached. Further increase of the delay time leads to an intensity growth for small clusters. However after this point the intensities of the medium size clusters decrease leading to that the spectra are dominated by small cluster sizes. In comparison to the corresponding spectra of (CO_2) (see figure (4.42)) CO shows higher fragmentation peaks.

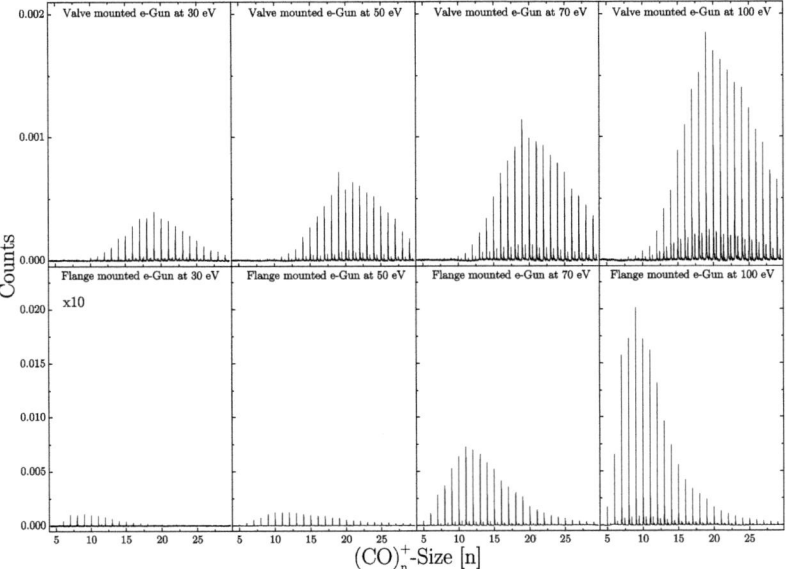

Figure 4.44 Depicted are the mass spectra for cluster sizes of $(CO)_5^+ - (CO)_{29}^+$. Comparison of the the valve mounted e-gun with the flange mounted e-gun for different ionization potentials. Molecular beam expansion of neat CO. Valve was held at 303 K and 4 MPa stagnation pressure. Both e-gun electrode potentials were optimized for maximal signal intensity at different ionization potentials. Filament current for the e-guns was 2.4 A. The mass spectra were recorded at 3 kV acceleration.

Influence of the ionization potential The influence of the e-gun filament potential ($\approx$electron energy) on the cluster size distribution is shown in a comparison of the valve mounted e-gun with the flange mounted e-gun for different applied filament potentials (note that the electron energy can differ from the applied filament potential due to space charge and other effects). With both setups the ion intensities increase with increasing ionization potential due to increase in electron current [261]. For the valve mounted e-gun the cluster size distribution does not shift with the ionization potential (upper row in figure 4.44). The high intensity of the magic number cluster $(CO)_{19}^+$ [106] is noticeable in all spectra recorded with the valve mounted e-gun. The intensity ratio of ca. 1 : 10 between the highest fragment peak and the corresponding parent cluster peak is independent from the ionization potential (upper row in figure 4.44). In the case of the flange mounted e-gun the ion intensities increase more significant with the ionization potential lower row in figure 4.44). Additionally the cluster size distribution shifts with increasing ionization potential to smaller clusters. The mass spectrum recorded at 100 eV ionization potential shows the characteristic log normal size distribution reported for cluster size distributions [101; 102]. These mass spectra for the flange mounted e-gun support the assumption that ionization after cluster formation increases the formation of smaller clusters by the fragmentation of bigger neutral clusters after ionization. In the case of the flange mounted e-gun the intensity ratio of ca. 1 : 24 between the highest fragment peak and the corresponding parent cluster peak does not change with ionization potential (lower row in figure 4.44).

Influence of the filament current The filament current of the flange mounted e-gun was gradually increased from the threshold of 2.1 A to 2.6 A (see figure 4.45). The potentials of the e-gun lenses were not modified. At the threshold current of 2.1 A and thus low electron emission the mass spectrum shows nearly no fragmentation. Increasing the filament current also increases space charge and thus the electron energy. Hence the count rate increase with the filament current and additionally first fragment peaks appear in the mass spectrum $((CO)_n^+C$ at 2.15 A in figure 4.45). At this point the maximum intensity for medium size clusters around $(CO)_{10}^+$ is reached. Further increases of the filament current do not increase the count rate of these cluster sizes. Increasing the filament current above 2.2 A dramatically increases the fragment amount in the mass spectra. Besides the fragment $(CO)_n^+C$ now also the fragment $(CO)_n^+O$ appear in the mass spectra and a general peak broadening with tailing peak bases can be observed. Again here the tailings of the peak bases indicate hot clusters which tend to decompose in the acceleration region of the mass spectrometer. This assumption is backed by the increase of the intensities of small clusters below $n \leq 6$ originating from the fragmentation of bigger clusters.

Influence of pulsed ionization Figure (4.46) shows the mass spectra of $(CO)_n^+$ cluster ions for pulsed operation of the flange mounted e-gun. The mass spectra show $(CO)_2^+ - (CO)_{19}^+$ clusters in dependence of the e-gun pulse duration. Up to

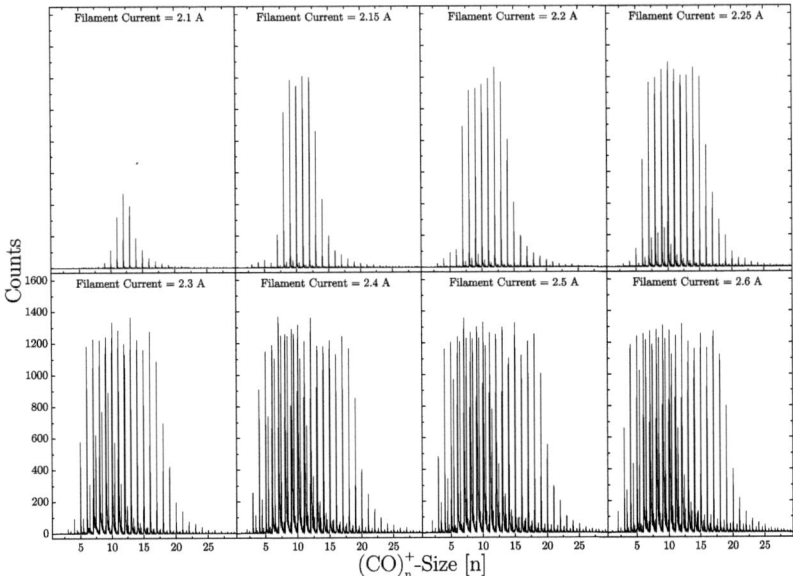

Figure 4.45 Depicted are the mass spectra of $(CO)_2^+ - (CO)_{30}^+$ clusters for different e-gun filament currents. Molecular beam expansion of neat CO. Valve was held at 305 K, 2.7 MPa stagnation pressure and 5 Hz repetition rate. The flange mounted e-gun was used at 250 eV electron energy. E-gun to valve distance was fixed at 20 mm. Mass spectra recorded at 3 kV acceleration and 2.5 μs extraction pulse for the Re-TOFMS. The extraction delay between valve opening and TOFMS extraction was fixed at 670 μs. Deflection plates were used bipolar with ±20 V deflection for increasing the intensity of small clusters.

1.5 μs increasing the pulse duration also increases the CO cluster ion yield. After reaching this threshold increasing the pulse length does not increase the count rate for the parent clusters. With longer pulsing times the peaks get broader and the fragment intensities grow (due to space charge and thus higher electron energy). A general peak broadening with tailing peak bases can be seen like in figure (4.45). With the increasing pulse time the space charge caused by the electrons and thus the Coulomb repulsion between them grow. This causes electron beam divergence which increases the interaction space between electrons and ions. So the possibility for ion molecule reactions will increase which would explain the increasing fragmentation observed in the mass spectra (see also figure 4.40).

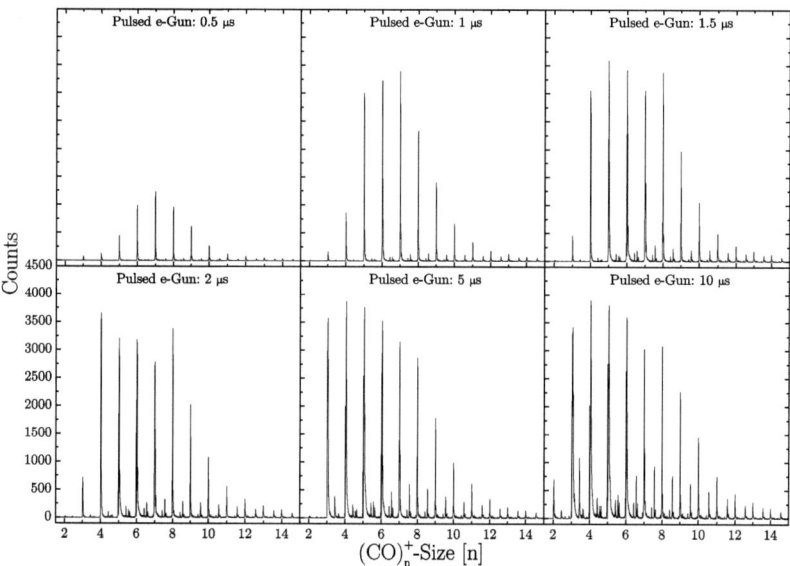

Figure 4.46 Depicted is the pulsed operation of the flange mounted e-gun. The mass spectra show $(CO)_2^+ - (CO)_{19}^+$ clusters in dependence of the pulse duration time. Molecular beam expansion of neat CO. Valve was held at 305 K and 2.5 MPa stagnation pressure. The flange mounted e-gun was used for ionization at 250 eV electron energy with 30 mm valve distance and 110 μs delay. Mass spectra recorded at 3 kV acceleration and 2.5 μs extraction pulse for the Re-TOFMS. The extraction delay between valve opening and TOFMS extraction was fixed at 665 μs. Deflection plates were used bipolar with ±25 V deflection for increasing the intensity of small clusters.

Influence of the deflector In figure (4.47) the upper row depicts the unfiltered mass spectra for $(CO)_3^+ - (CO)_{28}^+$ clusters in dependence of the deflection plates potentials. The cluster size distribution shifts with increasing plate potentials to smaller clusters. Fragmentation products grow with increasing plate potential and decrease after ±20 V. The lower row of figure (4.47) show the corresponding mass spectra of the size selected $(CO)_{10}^+$ cluster. Here also at first the intensity increases and reaches a maximum value for ±10 V. However few lighter mass peaks appear in the mass spectra showing the metastable decay of the parent cluster. Metastable decay of the parent is directly coupled to the fragment intensity seen in the upper row. Higher fragment yields indicate hotter clusters which also tend to more metastable decay. Here the question arises why different deflection plate potentials show different fragment ion yields? One possible explanation could be that for fixed extraction delays the variation of the deflection potential leads to the deflection of different portions of the molecular beam to the detector. The figure (4.43) show these different parts in dependence of the extraction delay time. With all these mentioned parameters (electron energy, nozzle to e-gun distance and so on) the cluster ion size distribution can be manipulated. In that sense one desired cluster size can be optimized in intensity for further scattering experiments. Therefore for total control of the cluster ion size and intensity two different electron guns are indispensable.

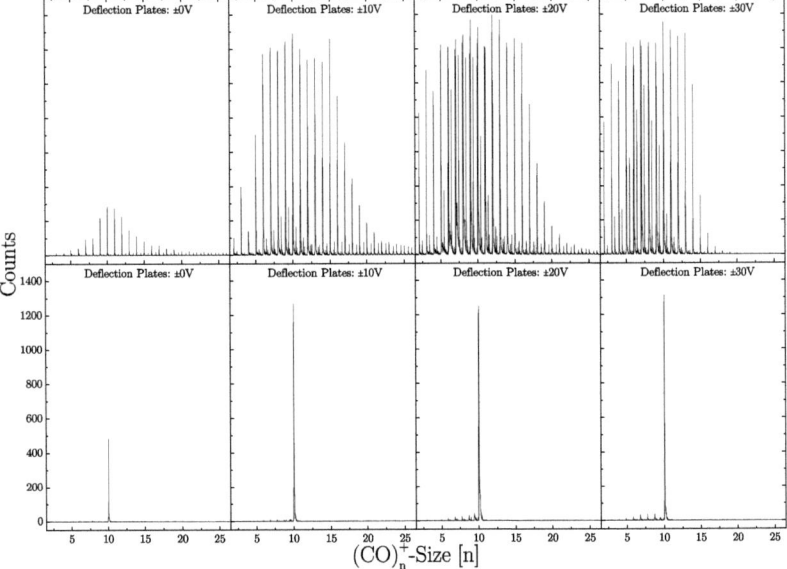

Figure 4.47 In the upper row the unfiltered mass spectra for $(CO)_3^+ - (CO)_{28}^+$ clusters in dependence of the deflection plates potentials are depicted. The lower row shows the corresponding mass spectra of the size selected $(CO)_{10}^+$ cluster. Molecular beam expansion of neat CO. Valve was held at 305 K and 2.9 MPa stagnation pressure. The flange mounted e-gun was used for ionization at 250 eV electron energy with 20 mm valve distance. Mass spectra recorded at 3 kV acceleration and 2.5 μs extraction pulse for the Re-TOFMS. The extraction delay between valve opening and TOFMS extraction was fixed at 630 μs. Deflection plates were used bipolar for increasing the intensity of small clusters.

4.3 Metastable Decay and Surface Impact

4.3.1 Impact of $(CO_2)_n^+$ on Stainless Steel Surface

In a series of surface impact experiments $(CO_2)_n^+$ clusters were impacted on the stainless steel backplane of the reflectron (last electrode). In order to decrease the mean cluster size and increase the intensity of small $(CO_2)_n^+$ clusters helium was used as seed gas (1:2 ratio) instead of argon. Prior to the impact experiments the intensity of the desired cluster size was maximized by variation of the stagnation conditions, ionization settings and other parameters described in the chapter before (see 4.2.5). The TOFMS accelerator potentials were set for optimum mass selection performance (space focus plane located at the mass gate). Besides this the reflectron is operated in reflection mode (surface potential higher than the TOFMS ion extraction potential). Before mass selection was applied the mass gate potentials were set to ground potential to avoid disturbance of the TOF spectra caused by charging of the gate wires. For mass selection the delay time between the ion extraction pulse and the mass gate transmission time window were adjusted for maximum transmission of a desired cluster size. Additionally the mass gate "opening" time window was narrowed to filter all other unwanted masses from the mass spectra up to the point where further decreasing of the gate "opening" time decreased the intensity of the selected cluster size. These two values were optimized to obtain clean "filtered" mass spectrum only with the mass peak of the desired sample cluster size. With these settings a reference TOF spectrum was recorded in reflection mode to which other TOF spectra and intensities of the same measurement (day) were related to. Due to the change in the intensity of the selected cluster size during the measurement (over the day) every hour or after a fixed number of recorded TOF spectra reference TOF spectra with the initial values were recorded in reflection mode. These reference spectra were used to plot the change in intensity (integrated peak area) for the selected cluster size over time. Due to the slow change in intensity two neighboring reference points were interpolated with linear functions. Hence the resulting linear fit functions were used to normalize the intensity of the measured TOF spectra which were recorded in between these two reference points. For similar expansion conditions the highest parent cluster ion signal intensity was observed for $(CO_2)_{14}^+$ (with 22700 counts/min) and the lowest parent cluster ion signal intensity was observed for $(CO)_{24}^+$ (with 2100 counts/min) respectively.

In conventional TOFMS apparatuses operated in reflection mode the potential barrier given by the reflection potential (U_{R2} in a two stage reflectron) is high enough to reflect the incoming ions (see figure 2.6 for reflectron principle and figure 4.31 for the potential energy view of the reflectron in reflection mode). In that case the reflected ions do not interact with the surface due to the insufficient penetration depth into the reflectron. The penetration depth of the ions into the reflectron depends on the ion kinetic energy and the reflection potential barrier given by U_{R2}. The cluster ions obey a kinetic energy distribution which is defined by the ion initial velocity distribution $f(v_0)$ (see equation 2.39) and the starting potential in the accelerator which is given by the initial starting position x_0 (see figure 2.5). Hence by successively decreasing the surface potential U_{R2} the penetration depth of the size selected cluster ions in the reflectron can be increased. In that case at first the ions with higher kinetic energy will impact the stainless steel backplane of the reflectron. With further decrease of the surface potential U_{R2} additionally the ions with smaller kinetic energy can penetrate deeper into the reflectron and interact with the surface, too. Thus the mean impact energy E_i is given by the difference between the mean ion kinetic energy (mean ion kinetic energy E_0 given by the spatial distribution in the TOFMS accelerator) and the potential energy of the surface [216; 281] in the form:

$$E_i = E_0 - eU_{R2} \qquad (4.3)$$

In the equation (4.3) only the perpendicular component to the surface $E_\perp$ of the ion velocity is regarded and the surface parallel component $E_\parallel$ [16] of the ion velocity is neglected (the perpendicular component $E_\perp$ is several orders of magnitude larger than the parallel component $E_\parallel$)[4]. The kinetic energy distribution and the mean kinetic energy E_0 of the ions can be measured by using the reflectron as a retarding field energy analyzer (surface with high neutralization efficiency is required e. g. stainless steel). Beginning with the reflection mode settings the potential of the second reflectron stage and thus surface potential U_{R2} was decreased successively in constant steps (e.g. $\Delta U_{R2} = 10$ V). For every potential step a mass spectrum was recorded for at least 3000 sweeps in ion counting mode which took ≈ 5 min. In the case when the parent ion signal intensity was optimized beforehand (ionization configuration, mass gate delay and so on) in general one sample cluster size impact measurement was measured over a day. The surface impact measurements for various cluster sizes of $(CO_2)_n^+$ and $(CO)_n^+$ impacted on stainless steel were carried out in the same manner. During the measurement all other reflectron TOFMS settings except the surface potential U_{R2} were left fixed. In surface collision mode the mass calibration changes with the surface potential value U_{R2}. For every recorded mass spectrum a new mass calibration is required. For the surface collision mode TOF spectra can be converted to mass spectra by the use of the known parent mass peak. With decreasing surface potential U_{R2} the TOF value of the parent cluster size peak shifts to longer TOF values. Using a simplified form of equation (4.1 with $b = 0$) the parameter a can be calculated with $a = (m/z)/t_{\text{peak}}^2$ for different surface potentials U_{R2}. With this a parameter the TOF spectra can be converted to mass spectra. Another possibility is to

---

[4]The parallel component of the impact velocity $v_\parallel$ scales with the sinus of the incidence angle α. For a geometrically estimated incidence angle $\alpha \approx 3°$ the velocity $v_\parallel$-component would be ~ 5 % of the incident velocity v_0. Note that the kinetic energy scales with the square of the velocity.

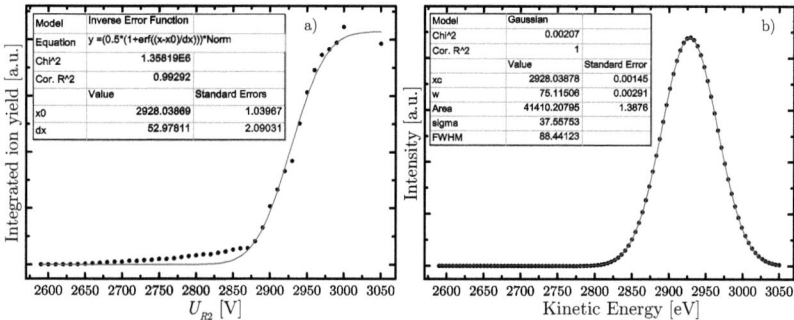

Figure 4.48 Collision of the CO_2^+ monomer with the stainless steel backplane of the reflectron collider. The reflectron TOFMS was operated in two stage mode at 3 kV extraction ($U_0 = 3$ kV, $U_1 = 2435$ V, $U_{R1} = 1647$ V, $\Delta U_{R2} = 10$ V and 300 eV EI) for CO_2 gas expansion seeded in helium (1:2 ratio, $P_0 = 1.7$ MPa and $T_0 = 305$ K). **a)** Depicted is the sigmoidal decrease in the integrated ion yield with decreasing surface potential U_{R2} (●). Here the reflectron collider is used as an energy analyzer assuming complete neutralization. The decrease in the integrated ion yield is well fitted with an error function (red solid line). **b)** Depicted is the first derivative of the error function of graph a (●) with a Gaussian shape. The red solid line is the Gaussian fit curve for the first derivative of the error function and represents the energy distribution with a mean kinetic energy $E_0 = 2928$ eV and with a FWHM of $\Delta E_{\text{FWHM}} = 88.4$ eV.

treat the TOFMS in collision mode as a TOF-TOF instrument as described in reference [16]. Figure (4.48 a) shows the result of a retarding field energy analysis for the surface impact of the CO_2^+ monomer on stainless steel. In figure (4.48) every point represents the integrated ion yield (counts) of the mass filtered CO_2^+ monomer peak spectrum recorded for different surface potentials U_{R2}. As already known from other retarding field energy measurements the decrease in the integrated ion yield shows sigmoidal behavior [16; 198; 200; 210]. By the way the decrease in the integrated ion yield is well fitted by an error function (red solid line in figure 4.48 a). However, a slight deviation from the error function fit curve is observed for $U_{R2} \leq 2875$ V. In that case the recorded integrated ion yield is higher than predicted by the error function fit curve and decreases nearly linearly with decreasing surface potential U_{R2}. This "tailing" behavior can be attributed to CO_2^+ ions scattered from the surface as was observed for the buckyball C_{60}^+ [252]. In the case of C_{60}^+ the signal of the scattered ions was enhanced by tilting the surface towards the detector. However, in our present setup the surface plane is parallel to the detector plane and cannot be tilted. The obtained error function in figure (4.48 a) was differentiated which is shown in figure (4.48 b, black points). First derivative of an error function is a Gaussian (due to the Gaussian origin of the error function which is an integral of a Gaussian distribution). Therefore the FWHM of the first derivative was determined by a Gaussian fit curve which is shown as a solid red line in figure (4.48 b). According to the fit curves (error function and Gaussian in figure 4.48 a and b) the mean kinetic energy of the monomer ions is $E_0 = 2928$ eV and the energy distribution is $\Delta E_{FWHM} = 88.4$ eV. For the impact of the monomer CO_2^+ ion no fragmentation products could be observed over the whole collision energy except of few ion counts around the mass 32 amu indicating a fragment corresponding to the dissociation product O_2 at higher collision energy (carbon atom loss with -12 amu, possibly EI dissociation product, see 4.2.5). However, the signal around mass 32 amu was very weak to be separated from the noise caused by the detector dark count rate (and could be therefore also an artifact). In the case of the dimer $(CO_2)_2^+$ this peak with the mass of the parent peak mass minus 12 amu $((CO_2)^+O_2)$ appears again in the mass spectra besides the parent peak.

Figure 4.49 schematically depicts the seven different regions of the TOFMS apparatus. Fragmentation of clusters by metastable decay can occur in all of these regions but is only detectable when it occurs in some of them which will be discussed below. In the first region the cluster ions are generated by EI (the main fragmentation reason). Depending on the ionization parameters the yield of fragments can increase or decrease which can be indirectly observed in the mass spectra (shift of the cluster size distribution as discussed in subsection 4.2.5). The second region depicted in figure (4.49) is the region after and between the skimmers. Depending on the valve to skimmer distance (and valve stagnation pressure) skimmer interference can occur which can increase fragment intensities and decrease parent cluster intensities. The hot clusters (heated up by EI and skimmer interference) cool down by metastable decay which influence mass spectra (cluster size distribution). Hence, subsequent peak broadening and tailing can be observed for metastable decay in the third region of the TOFMS apparatus, the accelerator. In that case the metastable origin of the ions can be checked by kinetic energy analysis. Metastable products generated in the TOFMS accelerator would possess a different kinetic energy according the time of birth and starting potential (even same mass). The fourth region is the space between the TOFMS accelerator and mass gate (quasi field free). Metastable decay which occurs in this region cannot be detected due to the fact of mass selection with the mass gate (except fragment ions with nearly the same TOF to the mass gate as the parent cluster ions). Only metastable decay products formed in the next field free region between the mass gate and reflectron are distinguishable without doubt (see equation (4.4) below). These decay products possess different kinetic energies and are separated in TOF due to different flight path and penetration depth in the reflectron (thus different resolution and peak width). According to these properties these ions show apparent masses. The apparent masses of these ions change with changes in the reflectron potentials (the change in the apparent mass is relative to the parent ion mass). Similar behavior is observed for the de-

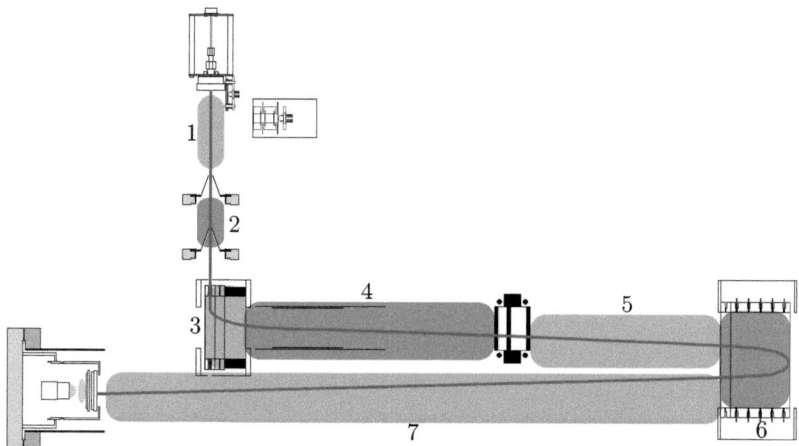

Figure 4.49 The figure depicts schematically the different regions of the TOFMS apparatus (see also figure 3.1) where metastable decay can be observed. 1. Main fragmentation region due to EI (shift of the cluster size distribution, see subsection 4.2.5); 2. Increased fragmentation due to skimmer interference (observable in the mass spectra due to fragmentation in the TOFMS accelerator); 3. Fragmentation of the "hot" clusters (generated by EI and skimmer interference) in the TOFMS accelerator (peak broadening and tailing, detectable by mass spectra and energy analysis); 4. Metastable decay in the "field free" region before mass selection (decay products will be filtered by the mass gate); 5. Metastable decay in the field free region after mass selection (detectable by energy analysis and TOF separation in the reflectron, decay products show an apparent mass); 6. Metastable decay in the reflectron or due to surface impact (not field free, detectable by energy analysis, decay products show an apparent mass); 7 Metastable decay in the field free region between the reflectron and the MCP-Detector is possible but not distinguishable.

cay products which are born in the non field free region of the TOFMS reflectron (region 6 in figure 4.49). These could be ions formed by metastable decay during deceleration, acceleration or surface impact in the reflectron. However due to the formation in a non field free region the mass of these ions cannot be derived from the kinetic energy of these ions by the utilization of equation (4.4) (the solution for surface impact induced fragments is described in subsection 4.3.3). The last region is the field free space between the reflectron exit mesh and the MCP-Detector. Although metastable decay can occur in this region it is not detectable due to the same TOF of the parent ions and metastable daughter ions. Here the intensity of the fragment peak is stronger than in the case of the impact of the monomer parent molecule. The resulting behavior of the integrated fragment ion

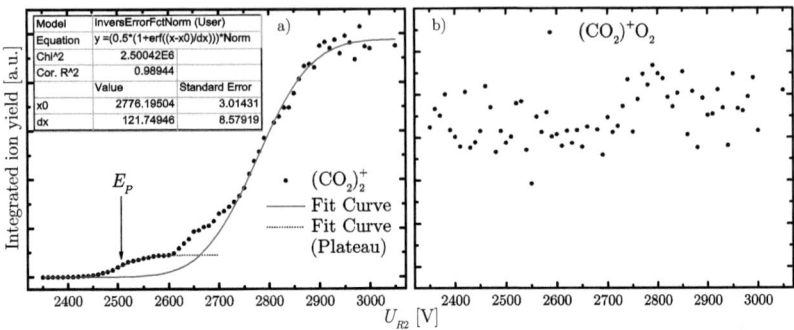

Figure 4.50 Collision of the dimer $(CO_2)_2^+$ with the stainless steel backplane of the reflectron collider. The reflectron TOFMS was operated in two stage mode at 3 kV extraction ($U_0 = 3$ kV, $U_1 = 2435$ V, $U_{R1} = 1647$ V, $\Delta U_{R2} = 10$ V and 200 eV EI) for CO_2 gas expansion seeded in helium (1:2 ratio, $P_0 = 1.9$ MPa and $T_0 = 305$ K). **a)** The sigmoidal decrease in the integrated ion yield with decreasing surface potential U_{R2} (•). Here the reflectron collider is used as an energy analyzer. The decrease in the integrated ion yield is well fitted with an error function fit curve (red solid line) except for surface potential values below $U_{R2} = 2730$ V. For $U_{R2} \leq 2730$ V first a steep decrease is observed which is followed by a intensity plateau with subsequent decrease to zero intensity (sigmoidal). **b)** Depicted is the integrated ion yield for the fragment with an apparent mass of $m_{(CO_2)^+O_2} = 76$ amu (•) which is reached at low surface potentials (e. g. $U_{R2} = 2400$ V). The integrated intensity is independent from the surface potential for the observed surface potential domain. The apparent mass and the integrated ion yield of the fragment indicate a metastable decay product.

peak intensity in dependence of the surface potential U_{R2} is depicted in figure (4.50 b). The integrated ion yield of the fragment peak is independent from the surface potential U_{R2} and is fluctuating around an integrated mean ion yield value. Another property of the fragment peak is that it shows an apparent[5] mass. The apparent mass of the fragment peak $(CO_2)^+O_2$ is for e. g. $U_{R2} = 2700$ V nearly 79 amu and reaches with decreasing U_{R2} the real value of 76 amu ($U_{R2} = 2400$ V, mass shift exemplary shown for the tetramer in figure 4.52). Both properties of the $(CO_2)^+O_2$ fragment peak, the apparent mass and no significant dependence on the collision energy indicate a metastable decay product. Contrary to this an impact induced fragment would show a collision energy and thus surface potential dependent intensity. An additional interesting feature is observed for the impact

---

[5]Metastable decay products possess a portion of the parent ion kinetic energy. In general this kinetic energy is not equal to the kinetic energy of an ion with the same mass but originating from the TOFMS accelerator. Hence metastable product ions will follow a different flight path and will show an "apparent" mass.

of the parent dimer $(CO_2)_2^+$ in the energy analysis plot (see figure 4.50 a). The data points are well fitted with the model fit function except for surface potential values below $U_{R2} = 2730$ V. In figure (4.50 a) first a slight decrease in the integrated ion yield is observed for $U_{R2} \leq 2730$ V which is followed by an intensity plateau with subsequently decreases to zero integrated ion yield with sigmoidal shape. The first slight decrease can be attributed to scattered parent ions as in the case of the monomer. However, the following sigmoidal decrease is an indication for a species with the same mass ($m_{(CO_2)^+} = 88$ amu) but with a different kinetic energy (birth potential). Stairs et al. observed in their energy analysis similar intensity plateaus for metastable fragment products of the zirconium Met-Car [282]. But in their study they used the reflectron in hard reflection mode (the ions are reflected in the first stage of the reflectron). In hard reflection mode the daughter ions and parent ions exhibit the same TOF and thus calculated mass. In the present study the reflectron is used in two stage mode as an energy analyzer (described as soft reflection mode in ref. [282]). In the case when the reflectron is operated in two stage mode the parent ions and daughter ions are separated in TOF and thus in mass too (due to different kinetic energies and penetration depth). Hence the intensity plateau shown in figure (4.50 a) cannot be ascribed to metastable decay of the parent dimer ion in the field free drift region but by decay in the acceleration region of the TOFMS. This intensity plateau was also well fitted by an error fit function which yields a mean kinetic energy of $E_P = 2504.5$ eV (see figure 4.50 a). Contrary to this the fit with an error function of the whole measurement in figure (4.50 a) yields a mean kinetic energy for the dimers with $E_0 = 2776.2$ eV. Regarding the relation between kinetic energy and mass ratio of the parent ion and daughter ion, the unknown mass m_d of a daughter ion can be calculated by the equation [170; 282–284]:

$$m_d = \left(\frac{U_d}{U_p}\right) m_p \qquad (4.4)$$

In equation (4.4) m_p is the mass of the parent ion, U_d the potential which defines the mean kinetic energy of the daughter ions $E_d = eU_d$ and U_p the birth potential of the parent ions with $E_0 = eU_p$. Equation (4.4) offers a method to proof the metastable decay origin of the ions causing the intensity plateau. Metastable decay products generated in the field free region would obey equation (4.4). In that case the equation (4.4) would yield "realistic" daughter ion masses m_d. By entering e. g. the values obtained for the intensity plateau in figure (4.50 a) with $U_p = 2776.2$ V, $m_p = 88$ amu and $U_d = 2504.5$ eV in equation (4.4) a "daughter" mass $m_d = 79.4$ amu is obtained. This would mean that the mass difference between the parent ion and daughter ion is $\Delta m = 8.6$ amu which is an "unrealistic" value. The calculation with equation (4.4) show that the intensity plateau observed in figure (4.50 a) cannot originate from the fragmentation of the dimer $(CO_2)_2^+$ by metastable decay in the field free region of the TOFMS.

An explanation for the occurrence of the intensity plateau could be metastable decay in the accelerator. In that case during the acceleration a bigger "hot" parent cluster looses dimers by metastable decay. These dimers start at different birth potentials U_0 than the dimers which are already present in the beam before the accelerator extraction pulse is applied. Thus these dimers have the same mass but a different birth potential and thus exhibit a different kinetic energy. Metastable decay in the acceleration region causes peak broadening and tailing in the TOF spectra. This peak broadening and tailing effect was observed for the dimer $(CO_2)_2^+$ TOF spectra for $U_{R2} \geq 2600$ V values above the intensity plateau (see also subsection 4.2.5). In the TOF spectra for the dimer impact besides the peak corresponding to the mass of $(CO_2)^+O_2$ an additional fragment with smaller mass appears too. Here again the fragment shows an apparent mass which tends the value of the monomer mass $m_{CO_2} = 44$ amu. However, the intensity of this peak was too small for further analysis. The same peak which corresponds to a monomer loss by metastable decay was also observed for the surface impact of the trimer $(CO_2)_3^+$ on the stainless steel reflectron backplane (see figure 4.51). In the case of the stainless steel surface impact of the trimer $(CO_2)_3^+$ both fragments, the fragment which corresponds to monomer loss ($(CO_2)_2^+$) and the fragment which corresponds to carbon loss ($(CO_2)_2^+O_2$) were observed in the TOF spectra. The resulting integrated ion yield of the impacted trimer $(CO_2)_3^+$ parent cluster and the integrated ion yield of the fragment ions $(CO_2)_2^+$ and $(CO_2)_2^+O_2$ are depicted in figure (4.51). The decrease of the integrated ion yield of the trimer is much steeper than in the case of the dimer (compare figure 4.50 a, and figure 4.51 a). Hence the deviation of the trimer integrated ion yield from the sigmoidal fit curve is smaller than in the case of the integrated ion yield of the dimer. For the trimer impact no clear intensity plateau is observed as in the case of the dimer impact. The deviation from the fit curve representing the ideal sigmoidal decrease of the integrated ion yield is more pronounced for surface potentials $U_{R2} \leq 2710$ V. Here the integrated ion yield decreases in two steps ($U_{R2} = 2710$ V and $U_{R2} = 2650$ V see figure 4.51 a)). In both steps the shape seems to be sigmoidal with no intensity plateau. This shape could be originating from the superposition of the scattered ions and the trimers formed by fragmentation in the acceleration region with a different birth potential. Regarding the mean kinetic energy, both the dimer $(CO_2)_2^+$ and the trimer $(CO_2)_3^+$ show nearly the same value around $E_0 = 2770$ eV. However, the FWHM of the kinetic energy of the dimer is twice as broad as the FWHM of the trimer (dimer: $\Delta E_{\text{FWHM}} = 203$ eV and trimer: $\Delta E_{\text{FWHM}} = 102$ eV obtained from Gaussian fits of the derivatives, similar to figure 4.48 b). This could be an indication for the process that more dimers are formed due to fragmentation of "hot" bigger clusters in the acceleration region and time window than trimers (maybe multiple decay of hot clusters to dimers). In the case of the fragment mass corresponding to $(CO_2)_2^+O_2$ (see figure 4.51 b) the integrated ion yield is more or less independent from the surface potential U_{R2} and is fluctuating around an integrated mean ion yield value (for the observed surface potential range). The

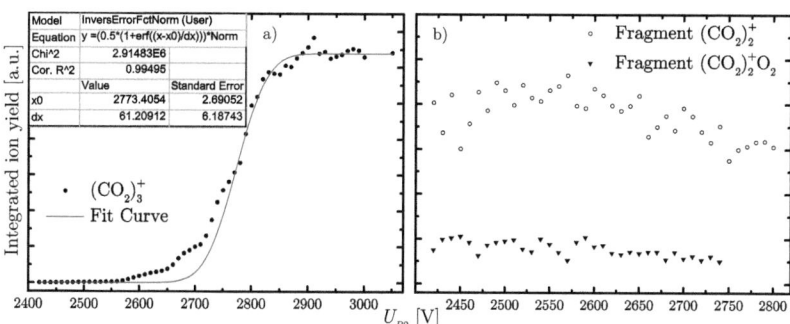

Figure 4.51 Collision of the trimer $(CO_2)_3^+$ with the stainless steel backplane of the reflectron collider. The reflectron TOFMS was operated in two stage mode at 3 kV extraction ($U_0 = 3$ kV, $U_1 = 2435$ V, $U_{R1} = 1647$ V and 200 eV EI with the flange mounted e-gun) for CO_2 gas expansion seeded in helium (1:2 ratio, $P_0 = 2.1$ MPa and $T_0 = 305$ K). **a)** The sigmoidal decrease in the integrated ion yield with decreasing surface potential U_{R2} (●). In that case the reflectron collider is used as an energy analyzer. The decrease in the integrated ion yield is well fitted with an error function fit curve (solid line) except for surface potential values below $U_{R2} = 2760$ V. For $U_{R2} \leq 2760$ V first a steep decrease is observed which is followed by two intensity plateaus (sigmoidal) with subsequent slight decrease to zero intensity. **b)** Depicted is the integrated ion yield for the two fragments corresponding to the mass of $m_{(CO_2)_2^+O_2} = 120$ amu (▼) and the mass of $m_{(CO_2)_2^+} = 88$ amu (○). Both fragments show apparent masses which are reached at low surface potentials (e. g. $U_{R2} = 2400$ V). For both fragments the integrated intensities are nearly independent from the surface potential U_{R2} for the observed surface potential domain. The apparent mass and the integrated ion yield of the fragment indicate a metastable decay product (see text).

other fragment mass corresponding to $(CO_2)_2^+$ which is formed by the loss of a monomer shows similar behavior. However, here the integrated ion yield shows a slight increase for lower potential values U_{R2}. This increase could be attributed to impact induced decay of the trimer though the increase is small and could be due to change in reflectron ion optical properties for low surface potentials U_{R2}. These fragments show also a surface potential U_{R2} dependent apparent mass. In the case of the fragment corresponding to the mass of $(CO_2)_2^+$ the change in the apparent mass is more pronounced than in the case of the fragment corresponding to the mass of $(CO_2)_2^+O_2$. Two representative mass spectra of the stainless steel surface impact of the trimer $(CO_2)_3^+$ are shown for two different surface potentials ($U_{R2} = 2550$ V and $U_{R2} = 2500$ V) in figure (4.52). In both mass spectra the mass is calibrated according to the TOF of the parent cluster $(CO_2)_3^+$ as described before. The fragment with a mass corresponding to carbon atom loss $(CO_2)_2^+O_2$

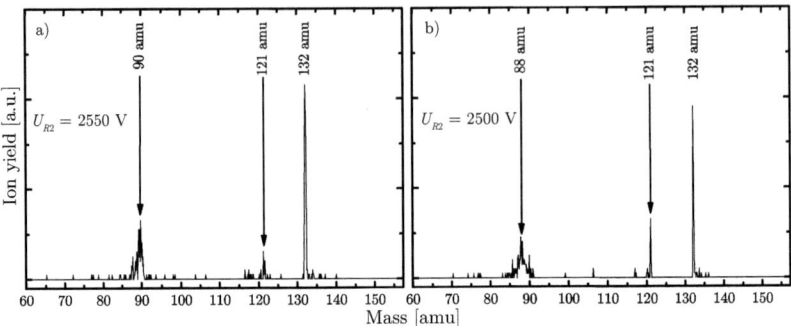

Figure 4.52 Collision of the trimer $(CO_2)_3^+$ with the stainless steel backplane of the reflectron collider. The reflectron TOFMS was operated in two stage mode at 3 kV extraction ($U_0 = 3$ kV, $U_1 = 2435$ V, $U_{R1} = 1647$ V and 200 eV EI with the flange mounted e-gun) for CO_2 gas expansion seeded in helium (1:2 ratio, $P_0 = 2.1$ MPa and $T_0 = 305$ K). **a)** Mass spectrum recorded in surface impact mode for the surface potential $U_{R2} = 2550$ V. Besides the parent cluster peak of the trimer $(CO_2)_3^+$ with mass 132 amu the fragment corresponding to carbon atom loss $((CO_2)_2^+ O_2)$ and the fragment corresponding to monomer loss $((CO_2)_2^+)$ are present in the mass spectra. The fragment formed by monomer loss shows here an apparent mass of 90 amu. In the case of the fragment formed by carbon loss the apparent mass tends to the exact value of 120 amu **b)** By further lowering of the surface voltage ($U_{R2} = 2500$ V) both fragment masses reach nearly the expected values. The apparent mass of the fragments and the integrated ion yield of the fragment indicate more or less the origin of the fragments as metastable decay products.

shows no significant change in the mass (around 121 amu) and shifts slowly with decreasing surface potential U_{R2} to the expected value of 120 amu. Compared to this the fragment with a mass corresponding to monomer loss $(CO_2)_2^+$ shows an apparent mass which changes faster with decreasing surface potential. The apparent mass shifts for $U_{R2} = 2550$ V from 90 amu to 88 amu for $U_{R2} = 2500$ V (see figure 4.52). Notable is also the peak shape of the fragment which corresponds to monomer loss. This peak has a much broader shape than the other fragment peak and the parent peak (mass resolution depends on the penetration depth in the reflectron which is different for metastable decay products). Due to the loss of a monomer the kinetic energy of the fragment ion and thus its penetration depth into the reflectron is decreased. Additionally to this kinetic energy is released by the fragmentation [166]. Both effects increase the peak width of the fragment mass peak. For the measurement of the cluster binding energy respectively the kinetic energy release peak shape analysis will be needed [166]. However, in order to obtain an accurate measurement of the ion peak shapes (including intensity and width), it is necessary to vary the reflectron potential U_{R2}, setting until the

4.3 Metastable Decay and Surface Impact

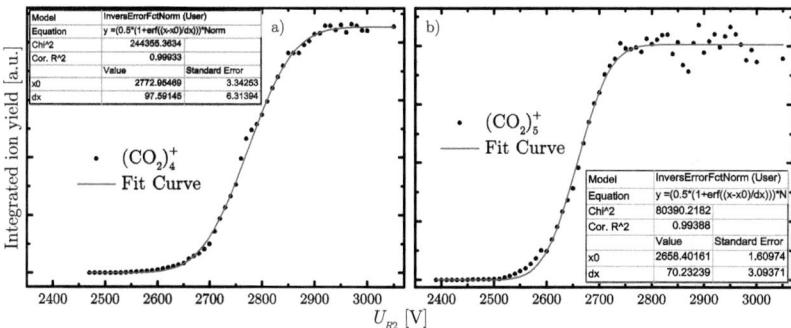

Figure 4.53 Collision of the tetramer $(CO_2)_4^+$ and pentamer $(CO_2)_5^+$ with the stainless steel backplane of the reflectron collider. The reflectron TOFMS was operated in two stage mode at 3 kV extraction ($U_0 = 3$ kV, $U_1 = 2435$ V, $U_{R1} = 1647$ V, 200 eV EI with the flange mounted e-gun of the tetramer and 300 eV EI with the valve mounted e-gun of the pentamer) for CO_2 gas expansion seeded in helium (1:2 ratio, $P_0 = 2.2$ MPa in the case of the tetramer and $P_0 = 2.7$ MPa in the case of the pentamer and $T_0 = 305$ K). **a)** The sigmoidal decrease in the integrated ion yield of the tetramer $(CO_2)_4^+$ with decreasing surface potential U_{R2} (•). Here the reflectron collider is used as an energy analyzer. The decrease in the integrated ion yield is well fitted with an error function fit curve (solid line) except small deviations for surface potential values around $U_{R2} = 2770$ V. No clear intensity plateau indicating fragmentation of bigger clusters to the tetramer is observed (200 eV EI with the flange mounted e-gun). **b)** The sigmoidal decrease in the integrated ion yield of the pentamer $(CO_2)_5^+$ with decreasing surface potential U_{R2} (•). The decrease in the integrated ion yield is well fitted with an error function fit curve (solid line) except a small intensity plateau located at $U_{R2} = 2600$ V indicating fragmentation of bigger clusters forming pentamers in the acceleration region (300 eV EI with the valve mounted e-gun).

parent and daughter ions have the same flight paths [167] which was not done in the present study. The apparent mass and peak shape are clear indications that these fragment clusters are formed by metastable decay in the field free region prior to the impact. Similar observations were made for the following cluster sizes, the tetramer $(CO_2)_4^+$ and pentamer $(CO_2)_5^+$. The results of the surface impact measurements for both cluster sizes are depicted in the figure (4.53). The sigmoidal decreases of the integrated ion yield of the parent clusters are well fitted with the error function curves. Compared to the other cluster sizes the tetramer $(CO_2)_4^+$ and the pentamer $(CO_2)_4^+$ (see figure 4.53 a) show less deviation from the "ideal" sigmoidal fit curve. An additional feature for the tetramer is that the graph lacks a distinct intensity plateau. Regarding the mean kinetic energy of the generated tetramer cluster ions, nearly the same value around $E_0 = 2770$ eV was obtained by the error fit function as in the case of the dimer and trimer. Figure (4.53 b) depicts the result of the surface impact measurement on stainless steel for the next bigger cluster size the pentamer $(CO_2)_5^+$. Here the sigmoidal decrease of the integrated parent ion yield is well fitted by the error fit function curve too. One clear difference compared to the tetramer graph is the small intensity plateau located around $U_{R2} = 2600$ V. Another difference compared to the dimer, trimer and tetramer is the value for the mean kinetic energy obtained by the error function fit. In the case of the pentamer the mean kinetic energy shifts to the value $E_0 = 2658$ V. An explanation for the different mean kinetic energies of the various cluster sizes could be the utilization of different ionization configurations. For the ionization of the monomer, dimer, trimer and tetramer the flange mounted e-gun was used with a relative distance of 30 mm to the valve head. The dimer, trimer, tetramer spectra were recorded with 200 V ionization potential and 2.5 A filament current. Thus these cluster sizes exhibit nearly the same mean ion kinetic energy around $E_0 = 2770$ V. To improve the monomer ion signal intensity the ionization potential was increased to 300 V and the filament current to 2.6 A. Due to the changes in the ionization settings the monomer exhibit quite a different mean ion kinetic energy with $E_0 = 2928$ V. The mass spectra for the clusters bigger than the tetramer were recorded with the valve mounted electron gun (300 V ionization potential and 2.4 A filament current). Thus for the pentamer the mean ion kinetic energy is located at $E_0 = 2658$ V (see figure 4.53). The electron gun settings affect electron beam geometry and space charge which can influence the mean ion kinetic energy. Hence for every scattering measurement the mean ion kinetic energy must be determined individually. For both cluster sizes the tetramer $(CO_2)_4^+$ and the pentamer $(CO_2)_5^+$ the same fragment products as in the case of the smaller clusters were observed. Both cluster sizes yield fragments with masses corresponding to monomer loss $(CO_2)_{n-1}^+$ and an additional oxygen molecule $(CO_2)_{n-1}^+ O_2$ in the cluster. The integrated yields of these fragment masses are depicted in figure (4.54) for the stainless steel surface impact of the tetramer (figure 4.54 a) and pentamer (figure 4.54 b). In general the integrated fragment ion yields for both cluster sizes colliding with the stainless steel surface are quite similar. A striking difference is that the integrated fragment ion yield for the surface impact of the tetramer is lower than in the case of the pentamer. This can be explained by the overall lower intensity of the mass selected tetramer parent cluster ion. For both colliding cluster sizes the integrated ion yield of the fragment with the mass corresponding to monomer loss $(CO_2)_{n-1}^+$ do not show any dependence on the applied surface potential U_{R2} (for the potential range of the measurements). Thus the fragment with the mass corresponding to monomer loss $(CO_2)_{n-1}^+$ seems to be a metastable decay product formed in the field free region of the TOFMS as in the case of the trimer. Regarding the other fragment with a mass corresponding to $(CO_2)_{n-1}^+ O_2$ in both graphs (figure 4.54 a and b), the integrated ion yields decrease nearly linearly with the surface potential U_{R2}. In the case of the tetramer (figure 4.54 a) the slope is smaller than in

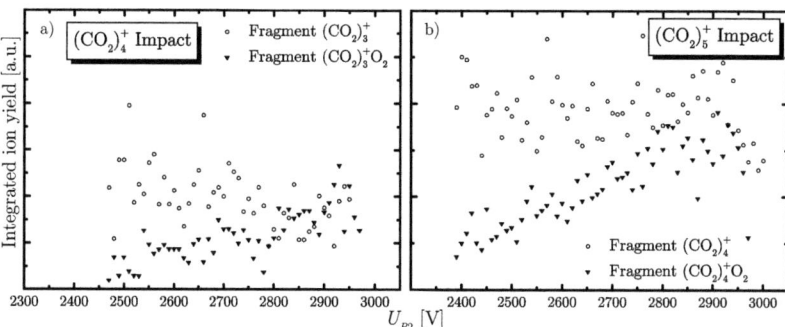

Figure 4.54 Integrated fragment ion yield for the surface collision measurement with the stainless steel backplane of the reflectron collider of the tetramer $(CO_2)_4^+$ and pentamer $(CO_2)_5^+$. The reflectron TOFMS was operated in two stage mode at 3 kV extraction ($U_0 = 3$ kV, $U_1 = 2435$ V, $U_{R1} = 1647$ V) for CO_2 gas expansion seeded in helium (1:2 ratio, $P_0 = 2.2$ MPa in the case of the tetramer and $P_0 = 2.7$ MPa in the case of the pentamer and $T_0 = 305$ K). **a)** Depicted is the integrated fragment ion yield for the fragments with the masses corresponding to $(CO_2)_3^+$ (○) and $(CO_2)_3^+O_2$ (▼, recorded at 200 V EI potential with the flange mounted e-gun). **b)** Depicted is the integrated fragment ion yield for the fragments with the masses corresponding to $(CO_2)_4^+$ (○) and $(CO_2)_4^+O_2$ (▼, recorded at 300 V EI potential with the valve mounted e-gun). Both integrated fragment ion yields are comparable to the integrated fragment ion yields of the tetramer impact (see graph a). Note the overall higher intensity of the fragments for the measurement of the pentamer (same scale, recorded at 300 V EI potential with the valve mounted e-gun).

the case of the pentamer (figure 4.54 b). Both graphs show no clear sigmoidal decrease in the integrated ion yield. Therefore it is difficult to distinguish the origin of these fragments. In the case of metastable decay these fragment ions would posses a portion of the kinetic energy of the parent cluster ions defined by the equation (4.4). Using the equation (4.4) the mean kinetic energy of the metastable fragments can be estimated. In the case of the pentamer e. g. the fragment corresponding to $(CO_2)_4^+O_2$ has a mass of $m_d = 208$ amu. Assuming that these fragments are formed by metastable decay of the parent pentamer ion in the field free drift region these ions would posses $E_d = 2513.4$ eV as their mean kinetic energy ($E_0 = 2658.4$ eV and $m_p = 220$ amu). However in the graph (4.54 b) there is no evidence for a sigmoidal decrease by the integrated fragment ion yield around the surface potential value of $U_{R2} = 2513$ V. A possible explanation for the origin of these fragments could be that these fragments are formed in the acceleration region by the decay of bigger clusters of the same series (e. g. $(CO_2)_n^+O_2 \Longrightarrow (CO_2)_{n-1}^+O_2 + CO_2$). The origin of these fragments is more clearer

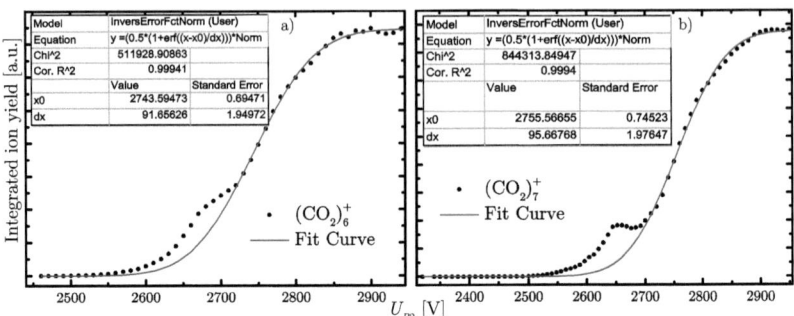

Figure 4.55 Collision of the hexamer $(CO_2)_6^+$ and heptamer $(CO_2)_7^+$ with the stainless steel backplane of the reflectron collider. The reflectron TOFMS was operated in two stage mode at 3 kV extraction ($U_0 = 3$ kV, $U_1 = 2435$ V, $U_{R1} = 1647$ V) for CO_2 gas expansion seeded in helium (1:2 ratio, $P_0 = 3.1$ MPa in the case of the hexamer and $P_0 = 3.3$ MPa in the case of the heptamer and $T_0 = 305$ K). **a)** Depicted is the sigmoidal decrease in the integrated ion yield of the hexamer $(CO_2)_6^+$ with decreasing surface potential U_{R2} (•). Here the reflectron collider is used as an energy analyzer. The decrease in the integrated ion yield is well fitted with an error function fit curve (solid line) except small deviations for surface potential values around $U_{R2} = 2700$ V. An intensity plateau indicating fragmentation of bigger clusters to the hexamer is observed (recorded at 300 V EI potential with the valve mounted e-gun). **b)** The sigmoidal decrease in the integrated ion yield of the heptamer $(CO_2)_7^+$ with decreasing surface potential U_{R2} (•). The decrease in the integrated ion yield is well fitted with an error function fit curve (solid line) except a distinct intensity plateau located around $U_{R2} = 2650$ V. According to the mean kinetic energy ratio between the main fit curve and the fit curve for the intensity plateau (not shown) the mean kinetic energy ratio indicate loss of 16 amu mass particle (recorded at 300 V EI potential with the valve mounted e-gun).

for the impact of the next cluster sizes, the hexamer and heptamer (note the use of the valve mounted e-gun). The stainless steel surface impact of the size selected hexamer $(CO_2)_6^+$ and heptamer $(CO_2)_7^+$ cluster ions are shown in figure (4.55). Compared to the tetramer $(CO_2)_4^+$ and pentamer $(CO_2)_5^+$ (figure 4.53) the hexamer and heptamer integrated ion yield curves show again intensity plateaus. The intensity plateau of the heptamer impact is much more pronounced than in the case of the hexamer. For both cluster sizes the intensity plateaus show clear sigmoidal behavior. To check the origin of these intensity plateaus the mean kinetic energies of these plateaus were determined by additional error fit functions (not shown). For the hexamer the mean kinetic energy obtained for the intensity plateau is located at $U_d = 2659$ V. Using equation (4.4) and $U_p = 2743.6$ V here again with $\Delta m = 8$ amu an unrealistic value for the mass difference is calculated indicating metastable decay in the acceleration stage as in the case of the dimer impact (see page 103). Regarding the intensity plateau of the heptamer, the mean kinetic energy is located at $U_d = 2612.5$ V and the mean kinetic energy according to the error fit function shown in figure 4.55 is located at $U_p = 2755.6$ V. Surprisingly with these values a daughter ion mass to parent ion mass difference of $\Delta m = 16$ amu is obtained using equation (4.4). The mass difference of $\Delta m = 16$ amu indicates metastable decay to $(CO_2)_6^+CO$ in the field free drift region which was not observed up to now. In a previous work [278] this kind of fragmentation was observed directly after ionization and not by decay in the field free region. Same results were obtained for the kinetic energy analysis of bigger clusters. In the case of e. g. the octamer $(CO_2)_8^+$ the mean kinetic energy of the parent cluster ions is $U_p = 2738.4$ V and the mean kinetic energy of the intensity plateau is localized around $U_d = 2614$ V (obtained by two different error fit function curves, not shown here). With these values a mass difference of $\Delta m = 16$ amu between the parent and daughter ion was calculated with equation (4.4) too. Additionally a different picture is observed for the integrated fragment ion yield of the hexamer and heptamer impact. The integrated ion yields for the hexamer impact and heptamer impact are shown in the figure (4.56). Significant different behavior is observed for the integrated fragment ion yield for the fragment with a mass corresponding to $(CO_2)_{n-1}^+O_2$. In both cases of the hexamer impact (figure (4.56) a) and the heptamer impact (figure (4.56) b) the integrated fragment ion yields show clear sigmoidal behavior compared to the smaller cluster sizes. Regarding the mean kinetic energy obtained by the error fit functions, in both cases the values are roughly the same as the mean kinetic energy of the parent clusters. This leads to the conclusion that these fragments obtained nearly the same mean kinetic energy in the acceleration region. The mean kinetic energy of these fragments (of the mass according to $(CO_2)_{n-1}^+O_2$) indicate that these fragments are already present in the ionized cluster beam formed by intracluster ion/molecule reactions [278]. Another difference to the smaller size clusters is that the integrated ion yields of the fragments according to $(CO_2)_{n-1}^+O_2$ are higher than the integrated fragment ion yields according to monomer loss $(CO_2)_{n-1}^+$ for higher U_{R2} values. In the case of the smaller clusters up to the hexamer the integrated fragment ion yield of the fragment according to monomer loss $(CO_2)_{n-1}^+$ was higher than the integrated fragment ion yield corresponding to $(CO_2)_{n-1}^+O_2$. Regarding the decrease of the integrated ion yield of the fragment corresponding to monomer loss $(CO_2)_7^+ \Longrightarrow (CO_2)_6^+ + CO_2$ (see figure 4.56 b), the curve shows a sigmoidal shape. However, it is a half sigmoidal curve due to the limited range of the measurement. Referring to the symmetry of the sigmoidal error fit function it was possible to fit the decrease of the integrated fragment ion yield corresponding to monomer loss to determine the mean kinetic energy of the fragment ions. In the case of the heptamer according to the error function fit curve the mean kinetic energy of the fragment corresponding to monomer loss is located at $U_d = 2392$ V. With $U_p = 2755.6$ V obtained by the fit curve (not shown) of the integrated ion

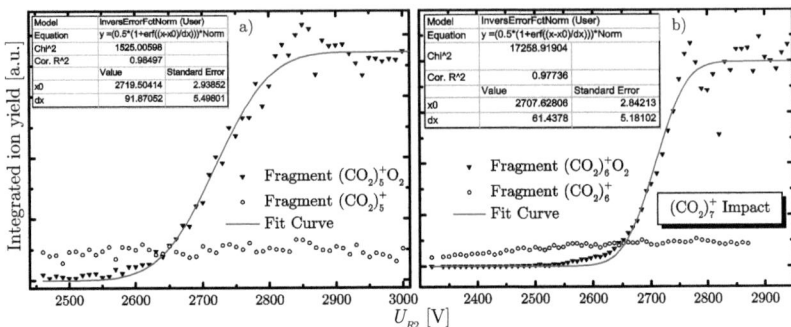

Figure 4.56 Integrated fragment ion yields for the collision of the hexamer $(CO_2)_6^+$ and heptamer $(CO_2)_7^+$ with the stainless steel backplane of the reflectron collider (see figure 4.55). **a)** Depicted is the integrated fragment ion yield for the fragments with masses corresponding to $(CO_2)_5^+O_2$ and $(CO_2)_5^+$ for the impact of the hexamer. The graph shows the sigmoidal decrease of the integrated fragment ion yield for the mass corresponding to $(CO_2)_5^+O_2$ (▼ with solid fit curve). The integrated fragment ion yield of the fragment corresponding to monomer loss ($(CO_2)_5^+$, ○) shows no dependence on the surface potential $U_{R2} \geq 2450$ V indicating metastable dissociation. **b)** Depicted is the integrated fragment ion yield for the fragments with masses corresponding to $(CO_2)_6^+O_2$ and $(CO_2)_6^+$ for the impact of the heptamer. The graph shows the sigmoidal decrease of the integrated ion yield of the integrated fragment ion yield for the mass corresponding to $(CO_2)_6^+O_2$ (▼ with solid fit curve). The integrated fragment ion yield of the fragment corresponding to monomer loss ($(CO_2)_6^+$, ○) shows sigmoidal behavior too, indicating metastable decay in the field free region (see text).

yield of the parent cluster the mass difference between the parent and daughter is 40.6 amu calculated with equation (4.4). Taking into account that this mass difference value is obtained by a "half" fit curve it confirms the metastable decay by monomer loss with 44 amu mass difference. Similar measurements with bigger size clusters verify that result.

4.3.2 Impact of $(CO)_n^+$ on Stainless Steel Surface

$(CO)_n^+$ clusters with $2 \leq n \leq 10$ In another series of experiments size selected carbon monoxide $(CO)_n^+$ cluster ions were impacted on the stainless steel backplane of the reflectron collider. The measurements were carried out under comparable experimental conditions described in the subsection (4.3.1). As in the case of carbon dioxide clusters the intensity of a desired cluster size was maximized by the optimization of the stagnation pressure, ionization settings and the other parameters described in the previous chapter (4.2.5). Small size $(CO)_n^+$ cluster ions were only observed with the flange mounted e-gun by increasing the distance between the valve and the electron source. Hence small cluster ions up to $(CO)_{12}^+$ were generated by the flange mounted e-gun at 150 V ionization potential to keep fragmentation at a low level. For similar expansion conditions the highest parent cluster ion signal intensity was observed for the carbon monoxide decamer $(CO)_{10}^+$ (with 46000 counts/min) and the lowest parent cluster ion signal intensity was observed for the carbon monoxide $(CO)_{25}^+$ (with 12000 counts/min) respectively. First attempts to use argon as a seed gas showed that carbon monoxide tends to form mixed clusters of argon and carbon monoxide. Therefore neat carbon monoxide gas was used for cluster formation. One main difference between the generated carbon dioxide cluster ions and the carbon monoxide cluster ions is that we not observed the carbon monoxide monomer cation. The monomer ion was absent in all carbon monoxide mass spectra even for different ionization and expansion conditions. This can be explained by the fact that smaller clusters are formed by the fragmentation of bigger clusters during EI. In that case the charge remains on the bigger fragment. Another fact is the stability of the carbon monoxide dimer $(CO)_2^+$ [13; 14] which favors the formation of dimer fragment ions. The kinetic energy analysis of the carbon monoxide dimer ion $(CO)_2^+$ is depicted in figure (4.57). In the case of the dimer the mean kinetic energy is located at $E_0 = 2791.7$ eV and shows a broad energy distribution of $\Delta E = 226.7$ eV (see figure 4.57 b). This indicates the formation of the dimer ions by fragmentation of bigger clusters in the acceleration region. The shape of the decrease in the integrated ion yield is clearly sigmoidal and well fitted with an error function except an intensity "shoulder" located around $U_{R2} = 2650$ V. The intensity "shoulder" indicates the formation of dimer ions by fragmentation of bigger clusters in the acceleration region, too. Besides this the intensity "shoulder" can also be the result of intact scattered dimer ions due to the less pronounced plateau. For the impact measurements of the dimer ions $(CO)_2^+$ no clear impact induced fragmentation or metastable fragmentation were observed. The shape of the integrated ion yield of the carbon monoxide dimer stainless steel impact is comparable with the carbon dioxide dimer stainless steel impact (see figure 4.50). Compared to the carbon monoxide dimer the following cluster size, the trimer, shows high fragment ion yields. Figure 4.58 shows the result for the stainless steel impact of the carbon monoxide trimer $(CO)_3^+$. In comparison to the carbon

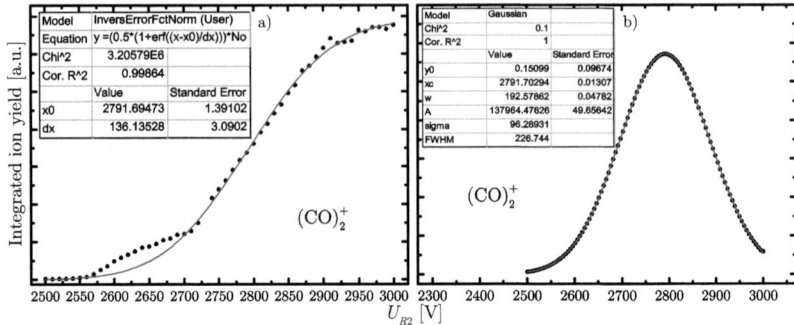

Figure 4.57 Collision of the dimer $(CO)_2^+$ with the stainless steel backplane of the reflectron collider. The reflectron TOFMS was operated in two stage mode at 3 kV extraction ($U_0 = 3$ kV, $U_1 = 2435$ V, $U_{R1} = 1647$ V and 150 eV EI with the flange mounted e-gun) for neat CO gas expansion ($P_0 = 1.7$ MPa and $T_0 = 305$ K). **a)** The sigmoidal decrease of the integrated ion yield with decreasing surface potential U_{R2} (•). Here the reflectron collider is used as an energy analyzer. The decrease in the integrated ion yield is except an intensity shoulder located around $U_{R2} = 2650$ V well fitted with an error function fit curve (solid line). **b)** Depicted is the first derivative of the error function fit curve of graph a) (•) with a Gaussian shape. The solid line is the Gaussian fit curve for the first derivative of the error function fit curve and represents the energy distribution with a mean kinetic energy $E_0 = 2791$ eV and with a FWHM of $\Delta E_{\text{FWHM}} = 226.7$ eV.

monoxide dimer and the carbon dioxide measurements the integrated ion yield of the carbon monoxide trimer shows no clear sigmoidal decrease (see figure 4.58 a). About three intensity shoulders are observed (for $U_{R2} = 2800$ V, $U_{R2} = 2750$ V and $U_{R2} = 2630$ V) in the graph. These intensity shoulders complicate finding a reliable error function fit. A possible error fit function for the sigmoidal decrease of the intensity of the integrated yield of the parent ion is depicted in figure (4.58 b) as a solid curve. The shape of the integrated ion yield of the parent carbon monoxide trimer ion indicates heavy fragmentation in the acceleration region. Hence many trimer ions are formed in the acceleration region by the fragmentation of bigger clusters. Another feature is the nearly linear decrease of the integrated ion yield below $U_{R2} = 2600$ V indicating intact scattered parent ions. This decrease is comparable to the decrease observed for the carbon monoxide dimer below $U_{R2} = 2650$ V. However in the case of the trimer the linearity of the decrease is much clearer pronounced. Regarding the fragment ion yields, two different fragments are observed. According to the mass spectra these fragments are identified as the dimer $(CO)_2^+$ formed by monomer loss and the dimer with a carbon atom $(CO)_2^+C$ formed by oxygen atom loss. The integrated fragment ion

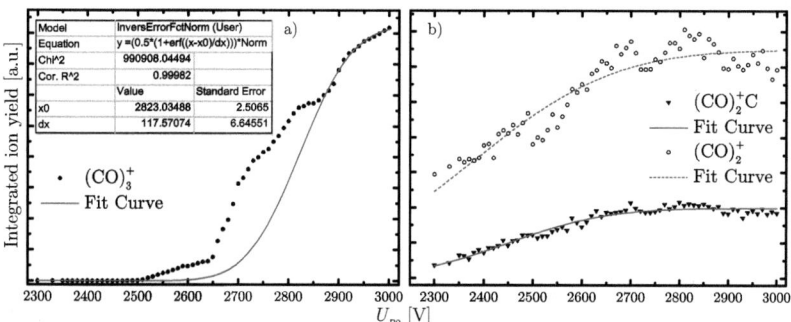

Figure 4.58 Collision of the trimer $(CO)_3^+$ with the stainless steel backplane of the reflectron collider. The reflectron TOFMS was operated in two stage mode at 3 kV extraction ($U_0 = 3$ kV, $U_1 = 2435$ V, $U_{R1} = 1647$ V and 150 eV EI with the flange mounted e-gun) for neat CO gas expansion ($P_0 = 1.7$ MPa and $T_0 = 305$ K). **a)** The decrease of the integrated ion yield with decreasing surface potential U_{R2} (•) with an estimated sigmoidal error function fit curve according to the shape (solid line). Here the reflectron collider is used as an energy analyzer. The decrease in the integrated ion yield shows several "shoulders" and a nearly linear decrease below $U_{R2} = 2600$ V indicating scattered trimers. **b)** Depicted are the integrated fragment ion yields of the fragment $(CO)_2^+C$ (▼) and the fragment $(CO)_2^+$ (○) observed for the impact of the trimer. Both fragment ion yields show sigmoidal decrease and are well fitted by error function fit curves (solid and dashed lines).

yields of both fragments show sigmoidal behavior and are well fitted with error function fit curves (see figure 4.58 b). Both fragments show high intensities and are already present in the mass spectra before the parent ions are colliding with the stainless steel surface. This fact excludes the formation of these fragments by surface impact induced fragmentation. Analysis of the mean kinetic energy of these fragments shows that these fragments are not formed in the field free region of the TOFMS. The mean kinetic energy of these fragments is higher than it would be expected for fragments with these masses formed in the field free region of the TOFMS by metastable decay. This indicates that these fragments are formed in the acceleration region of the TOFMS by the fragmentation of bigger clusters of the same series e. g. $(CO)_n^+$ and $(CO)_n^+C$. According to this assumption the lifetime of these clusters must be lower than the acceleration time of several micro seconds in the TOFMS accelerator. Similar results were obtained for bigger carbon monoxide clusters. In the case of the next cluster size, the tetramer $(CO)_4^+$, the integrated parent ion yield shows again deviations from the sigmoidal shape. Fitting of the data points with an error function is complicated by many intensity shoulders. Figure (4.59 a) shows the decrease of the integrated

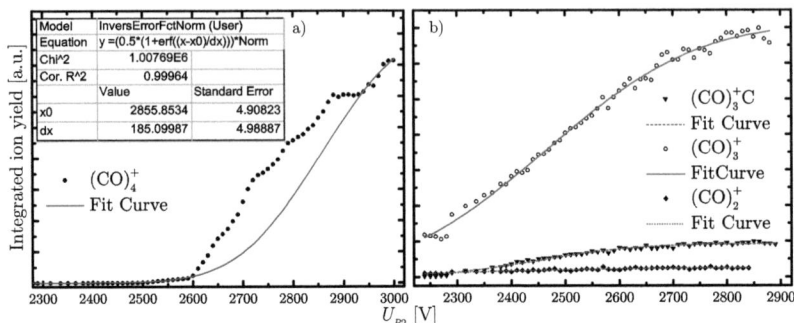

Figure 4.59 Collision of the tetramer $(CO)_4^+$ with the stainless steel backplane of the reflectron collider. The reflectron TOFMS was operated in two stage mode at 3 kV extraction ($U_0 = 3$ kV, $U_1 = 2435$ V, $U_{R1} = 1647$ V and 150 eV EI with the flange mounted e-gun) for neat CO gas expansion ($P_0 = 1.7$ MPa and $T_0 = 305$ K). **a)** The decrease of the integrated ion yield with decreasing surface potential U_{R2} (•) with an estimated sigmoidal error function fit curve according to the shape (solid line). Here the reflectron collider is used as an energy analyzer. The decrease in the integrated ion yield shows several "shoulders" and a nearly linear decrease below $U_{R2} = 2600$ V indicating intact scattered tetramers. **b)** Depicted are the integrated fragment ion yields of the fragment $(CO)_3^+C$ (▼), $(CO)_3^+$ (○) and the fragment $(CO)_2^+$ (♦) observed for the impact of the tetramer. All three fragment ion yields show sigmoidal decrease and are well fitted by error function fit curves (solid, dashed and dotted, see text for mean kinetic energy values).

parent ion yield with decreasing surface potential. The fit curve is based on the fit curve used for the trimer impact (and the bigger clusters e. g. the heptamer see figure 4.60). In that case the mean kinetic energy is given by $E_0 = 2855$ V, a little bit higher than the mean kinetic energy of the trimer with $E_0 = 2823$ V. Regarding the fragments, for the impact of the carbon monoxide tetramer three different fragments were observed: $(CO)_3^+C$, $(CO)_3^+$ and $(CO)_2^+$. The integrated fragment ion yields are depicted in dependence of the surface potential U_{R2} in figure (4.59 b). All integrated fragment ion yields decrease with decreasing surface potential U_{R2} and have sigmoidal shapes. Therefore all integrated fragment ion yield data points are well fitted with error function fit curves (see figure 4.59 b, $(CO)_3^+$: solid, $(CO)_3^+C$: dashed and $(CO)_3^+$: dotted lines). Only the fragment corresponding to $(CO)_3^+C$ has a mean kinetic energy indicating metastable decay in the field free region of the TOFMS. The mean kinetic energy of the fragment corresponding to $(CO)_3^+C$ is $E_d = 2451.2$ eV. According to equation (4.4) the daughter is formed by the loss of $\Delta m = 16$ amu corresponding to loss of an oxygen atom. In the case of the fragments corresponding to monomer loss the $(CO)_3^+$ and $(CO)_2^+$ the mean kinetic energy is higher than it would be expected for met-

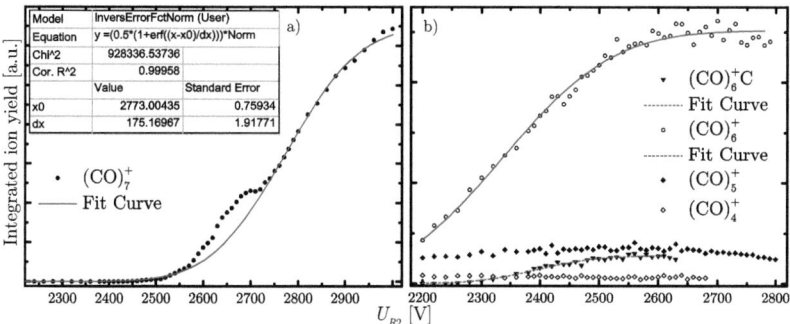

Figure 4.60 Collision of the heptamer $(CO)_7^+$ with the stainless steel backplane of the reflectron collider. The reflectron TOFMS was operated in two stage mode at 3 kV extraction ($U_0 = 3$ kV, $U_1 = 2435$ V, $U_{R1} = 1647$ V and 150 eV EI with the flange mounted e-gun) for neat CO gas expansion ($P_0 = 1.7$ MPa and $T_0 = 305$ K). **a)** The decrease of the integrated ion yield with decreasing surface potential U_{R2} (•) with an estimated sigmoidal error function fit curve according to the shape (solid line). Here the reflectron collider is used as an energy analyzer. The decrease in the integrated ion yield shows one distinct intensity plateau (around $U_{R2} = 2690$ V) and a nearly linear decrease below $U_{R2} = 2570$ V indicating scattered heptamers. **b)** Depicted are the integrated fragment ion yields of the fragment $(CO)_6^+C$ (▼), $(CO)_6^+$ (○), $(CO)_5^+$ (♦) and the fragment $(CO)_4^+$ (◇) observed for the impact of the heptamer. The two fragment ion yields of the $(CO)_6^+C$ and $(CO)_6^+$ show both sigmoidal decrease and are well fitted by error function fit curves (solid, dashed). The fragment ion yield of the fragment $(CO)_5^+$ show a slight increase with decreasing surface potential a possible indication for impact induced fragmentation (see text).

astable decay in the field free region. The mean kinetic energy of the fragment corresponding to $(CO)_3^+$ is $E_d = 2463.7$ eV and of the fragment corresponding to $(CO)_2^+$ is $E_d = 2302.4$ eV. Both fragments show apparent masses which converge with decreasing surface potential to the values of the dimer and trimer (56 amu and 84 amu). According to the apparent mass and the mean kinetic energy these fragments are formed in the acceleration region of the TOFMS accelerator or in the reflectron after the turning point during acceleration in the reflectron. Hence only the origin of the fragments formed in the field free region e. g. the fragment $(CO)_3^+C$ can be determined without doubts. Similar results were obtained for the impact of bigger clusters. The number of fragments x observed in the mass spectra corresponding to monomer loss $(CO)_{n-x}^+$ increases with the cluster size n. This increase in the number of fragments can be explained by the increasing experimental observation time window which increases with the TOF of the sample. Exemplary the stainless steel surface impact of the carbon monoxide heptamer is shown in the figure (4.60). With increasing cluster size the integrated parent ion yield shows less deviations from the sigmoidal shape (see figure 4.60 a). For the heptamer parent ion yield only one clear intensity plateau (around $U_{R2} = 2690$ V) is observed. Regarding the observed fragments, only the fragments corresponding to monomer loss $(CO)_6^+$ and oxygen loss $(CO)_6^+C$ show sigmoidal shapes with decreasing surface potential U_{R2} (see figure 4.60 b). However in both cases the mean kinetic energies do not match the kinetic energies that would be expected for the formation of these fragments in the field free region (as in the case of the tetramer impact). Therefore the formation of these fragments in the acceleration region of the TOFMS accelerator or in the reflectron prior or after the impact cannot be excluded. Another interesting feature is observed for the fragment corresponding to dimer loss corresponding to $(CO)_5^+$. The integrated fragment ion yield of the fragment $(CO)_5^+$ increases slightly with increasing collision energy and decreases again below $U_{R2} = 2400$ V. This behavior could be an indication for the superposition of two fragment formation effects, $(CO)_5^+$ fragment formation by metastable decay of the parent ion or by impact induced fragmentation of the parent ion. With the experimental setup used for this measurement it is not possible to distinguish between these two effects. Without the increase in the integrated fragment ion yield both fragments the $(CO)_5^+$ and $(CO)_4^+$ indicate that these fragments are formed by metastable decay. The observed picture does not change for the impact of the following cluster sizes, e. g. the nonamer $(CO)_9^+$ and decamer $(CO)_{10}^+$. Figure 4.61 shows the stainless steel surface impact of the nonamer $(CO)_9^+$. The mean kinetic energy $E_0 = 2765.2$ V of the nonamer parent ion (see figure 4.61 a) is in the range of the kinetic energy obtained for the heptamer ($E_0 = 2773$ eV, see figure 4.60 a). In the case of the nonamer impact the integrated ion yield exhibits again an intensity shoulder which is less pronounced than in the case of the heptamer impact. With increasing cluster size the shape of the intensity shoulders and plateaus gets less pronounced indicating that formation of these clusters by fragmentation in the acceleration region decreases. For the stainless steel surface impact of the nonamer parent six kind of fragments were observed: $(CO)_8^+C$, $(CO)_8^+$, $(CO)_7^+C$, $(CO)_7^+$, $(CO)_6^+$ and $(CO)_5^+$. The integrated fragment ion yields of these fragments except the $(CO)_7^+C$ fragment are depicted in figure (4.61 b). The integrated fragment ion yield of the $(CO)_7^+C$ fragment ion is not included to the graph due to low intensity. Three of the integrated ion yields ($(CO)_8^+C$, $(CO)_8^+$ and $(CO)_7^+$) show sigmoidal behavior and are all well fitted by error function fit curves (see figure 4.61 b, solid, dashed and dotted curves). Only in the case of the fragment $(CO)_8^+$ the mean kinetic energy obtained by the fitting procedure matches the expected mass difference of $\Delta m = 28$ amu (with $E_d = 2471$ eV). In the case of the two other fragments the $(CO)_8^+C$ and $(CO)_7^+$ the calculated mass differences according to the mean kinetic energies of the fragments do not match the expected mass difference. In the case of the fragment corresponding to $(CO)_8^+C$ the calculated mass difference is $\Delta m = 32$ amu ($E_d = 2423.5$ eV) indicating a lower mean kinetic energy than expected. Regarding the other fragment with a mass corresponding to $(CO)_7^+$, the

4.3 Metastable Decay and Surface Impact

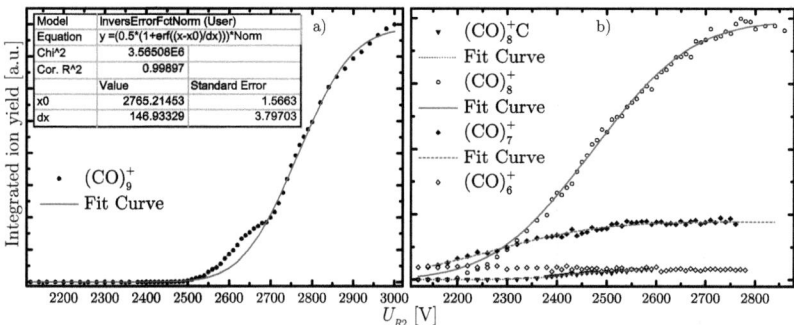

Figure 4.61 Collision of the nonamer $(CO)_9^+$ with the stainless steel backplane of the reflectron collider. The reflectron TOFMS was operated in two stage mode at 3 kV extraction ($U_0 = 3$ kV, $U_1 = 2435$ V, $U_{R1} = 1647$ V and 150 eV EI with the flange mounted e-gun) for neat CO gas expansion ($P_0 = 1.7$ MPa and $T_0 = 305$ K). **a)** The decrease of the integrated ion yield with decreasing surface potential U_{R2} (•) with an estimated sigmoidal error function fit curve according to the shape (solid line). Here the reflectron collider is used as an energy analyzer. The decrease in the integrated ion yield shows one distinct intensity plateau (around $U_{R2} = 2670$ V). **b)** Depicted are the integrated fragment ion yields of the fragment $(CO)_8^+C$ (▼), $(CO)_8^+$ (○), $(CO)_7^+$ (◆) and the fragment $(CO)_6^+$ (◇) observed for the impact of the nonamer. The fragment ion yields of the fragments $(CO)_8^+C$, $(CO)_8^+$ and $(CO)_7^+$ all show sigmoidal decrease and are well fitted by error function fit curves (solid, dashed and dotted). The fragment ion yield of the fragment $(CO)_6^+$ shows a slight increase with decreasing surface potential a possible indication for impact induced fragmentation (see also heptamer impact 4.60).

calculated mass difference is $\Delta m = 45$ amu ($E_d = 2271.1$ eV) indicating a higher mean kinetic energy than expected. According to these mass differences both fragments should be formed in the reflectron during deceleration or acceleration. Regarding the lightest fragment corresponding to $(CO)_6^+$, the integrated fragment ion yield shows a slight increase with decreasing surface potential, a possible indication for impact induced fragment generation. Figure 4.62 shows the stainless steel surface impact of the decamer $(CO)_{10}^+$ parent ion. The mean kinetic energy $E_0 = 2763.2$ V of the nonamer parent ion (see figure 4.62 a) is in the range of the kinetic energy obtained for the heptamer ($E_0 = 2773$ eV, see figure 4.60 a) and the kinetic energy obtained for the nonamer parent ion ($E_0 = 2765.2$ eV, see figure 4.61 a). In the case of the decamer $(CO)_{10}^+$ ion impact the integrated ion yield exhibits again an intensity shoulder which is less pronounced than in the case of the heptamer impact and nonamer impact. For the stainless steel surface impact of the decamer parent ion seven kind of fragment ions were observed: $(CO)_9^+C$, $(CO)_9^+$, $(CO)_8^+C$, $(CO)_8^+$, $(CO)_7^+C$, $(CO)_7^+$ and $(CO)_6^+$. The

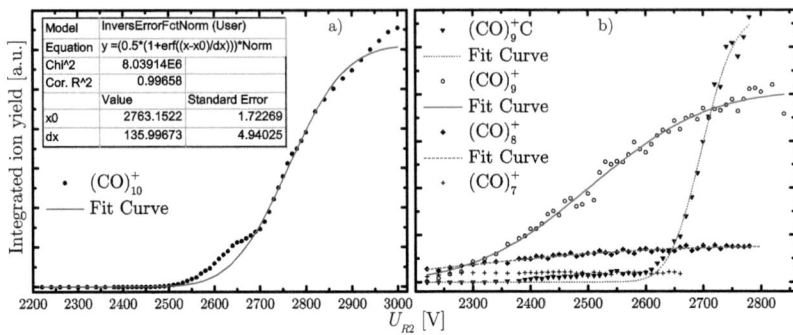

Figure 4.62 Collision of the decamer $(CO)_{10}^+$ with the stainless steel backplane of the reflectron collider. The reflectron TOFMS was operated in two stage mode at 3 kV extraction ($U_0 = 3$ kV, $U_1 = 2435$ V, $U_{R1} = 1647$ V and 150 eV EI with the flange mounted e-gun) for neat CO gas expansion ($P_0 = 1.7$ MPa and $T_0 = 305$ K). **a)** The decrease of the integrated ion yield with decreasing surface potential U_{R2} (•) with an estimated sigmoidal error function fit curve according to the shape (solid line). Here the reflectron collider is used as an energy analyzer. The decrease in the integrated ion yield shows one distinct intensity shoulder (around $U_{R2} = 2670$ V) comparable with the intensity shoulder observed for the nonamer impact (see figure 4.61 a). **b)** Depicted are the integrated fragment ion yields of the fragment $(CO)_9^+C$ (▼), $(CO)_9^+$ (○), $(CO)_8^+$ (◆) and the fragment $(CO)_7^+$ (+) observed for the impact of the decamer. The fragment ion yields of the fragments $(CO)_9^+C$, $(CO)_9^+$ and $(CO)_8^+$ all show sigmoidal decrease and are well fitted by error function fit curves (solid, dashed and dotted). The fragment ion yield of the fragment $(CO)_7^+$ shows a slight increase with decreasing surface potential a possible indication for impact induced fragmentation (see also heptamer impact 4.60) and nonamer impact 4.61.

integrated fragment ion yields of these fragments except the $(CO)_8^+C$, $(CO)_7^+C$ and $(CO)_6^+$ fragments are depicted in figure (4.62 b). The integrated fragment ion yields of the $(CO)_8^+C$, $(CO)_7^+C$ and $(CO)_6^+$ fragment ions are not included in the graph (4.62 b) due to low intensity. Three of the integrated ion yields ($(CO)_9^+C$, $(CO)_9^+$ and $(CO)_8^+$) show sigmoidal behavior and are all well fitted by error function fit curves (see figure 4.62 b, solid, dashed and dotted curves). Only in the case of the fragment $(CO)_9^+$ the mean kinetic energy obtained by the fitting procedure matches nearly the expected mass difference of $\Delta m = 28$ amu (with $E_d = 2504.2$ eV $\longrightarrow \Delta m = 26.3$ amu). In the case of the two other fragments the $(CO)_9^+C$ and $(CO)_8^+$ the calculated mass differences according to the mean kinetic energies of the fragments do not match the expected mass differences. Regarding the fragment corresponding to $(CO)_9^+C$, the calculated mass difference is $\Delta m = 6.7$ amu ($E_d = 2697.3$ eV) indicating a higher mean kinetic energy than expected. Regarding the other fragment with a mass corresponding to $(CO)_8^+$, the calculated mass difference is $\Delta m = 46$ amu ($E_d = 2309.3$ eV) indicating a higher mean kinetic energy than expected. According to these mass differences both fragments should be formed in the reflectron during deceleration or acceleration. Regarding the lightest fragment corresponding to $(CO)_7^+$, the integrated fragment ion yield shows a slight increase with decreasing surface potential, a possible indication for impact induced fragment generation. However, compared to the intensity increase observed for the $(CO)_5^+$ fragment ion of the heptamer impact and the $(CO)_6^+$ fragment ion of the nonamer impact the increase in the integrated fragment ion yield of the $(CO)_7^+$ fragment ion of the decamer impact is less pronounced and negligible. An additional interesting feature is observed for the heaviest fragment corresponding to $(CO)_9^+C$. Compared to the impact of the smaller clusters (e. g. the heptamer and nonamer) the mean kinetic energy of this fragment ion ($E_d = 2697.3$ eV) is close to the mean kinetic energy of the parent nonamer cluster ion ($E_0 = 2763.2$ eV) and less broad (compare the figures 4.60 b, and 4.61 b with figure 4.62 b).

$(CO)_n^+$ clusters with $25 \leq n \leq 30$ As described before (see subsection 4.2.5) the cluster size distribution is mainly influenced by the ionization source and conditions. Therefore bigger size carbon monoxide cluster ions (with $n \geq 25$) were generated with the valve mounted e-gun. These clusters were also mass selected and used for stainless steel surface collision experiments. Compared to the smaller size clusters (with $n \leq 10$) which were generated by the flange mounted e-gun the fragment ion intensities observed for the bigger size clusters which were generated with the valve mounted e-gun are relatively small. Therefore in most cases the fragments corresponding to the series $(CO)_{n-x}^+C$ (with $x = 1, 2$ and 3) were observed but the ion yield of these fragments were too low for further analysis. Similarly low fragment ion yields were observed for fragments of the series corresponding to $(CO)_{n-x}^+$ (with $x \geq 4$). These fragments were also excluded from analysis and are not shown in the graphs. The main difference of the bigger cluster impact measurement compared to the smaller cluster impact measurement is that clear indications for impact induced fragmentation could be observed. This will be discussed on some representative bigger cluster impact measurements beginning with the $(CO)_{25}^+$ impact. The results obtained for the impact measurement of the $(CO)_{25}^+$ parent cluster ion are shown in the figure (4.63). Again the integrated parent cluster ion yield shows sigmoidal behavior with decreasing surface potential (see figure 4.63 a). The decrease of the integrated parent cluster ion yield is well fitted with an error function fit curve except an intensity shoulder located around $U_{R2} = 2650$ V. With the fit curve a mean kinetic energy of $E_0 = 2722$ eV is determined for the parent cluster ions. Compared to the smaller cluster ions the intensity shoulder has a "tailing" shape without any intensity plateau. Noticeable are some intensity variations of the integrated parent ion yield e. g. at $U_{R2} = 2560$ V (see figure 4.63 a). The increase of the

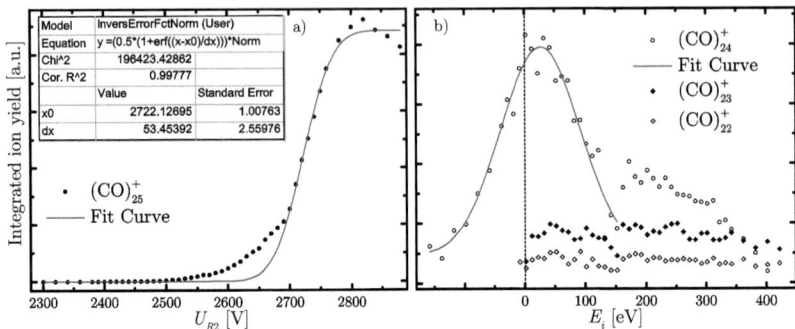

Figure 4.63 Collision of the $(CO)_{25}^+$ parent cluster ion with the stainless steel backplane of the reflectron collider. The reflectron TOFMS was operated in two stage mode at 3 kV extraction ($U_0 = 3$ kV, $U_1 = 2435$ V, $U_{R1} = 1647$ V and 300 eV EI with the valve mounted e-gun) for neat CO gas expansion ($P_0 = 3.5$ MPa and $T_0 = 305$ K). **a)** The decrease of the integrated ion yield with decreasing surface potential U_{R2} (•) with a sigmoidal error function fit curve (solid line). Here the reflectron collider is used as an energy analyzer. The decrease in the integrated ion yield shows one slight intensity shoulder (around $U_{R2} = 2650$ V) compared to the smaller size clusters with $n \leq 10$. **b)** Depicted are the integrated fragment ion yields of the fragment $(CO)_{24}^+$ (○), $(CO)_{23}^+$ (◆) and the fragment $(CO)_{22}^+$ (◇) observed for the impact of the $(CO)_{25}^+$ parent cluster ion. None of the fragment ion yields show sigmoidal decrease. The fragment ion yield of the fragment $(CO)_{24}^+$ shows a nearly symmetrical and slightly shifted peak around the zero collision energy line ($E_i = 0$ eV), an indication for impact induced fragmentation. Hence, the integrated ion yield of the fragment corresponding to $(CO)_{24}^+$ is well fitted by a Gaussian fit curve. The center of the Gaussian fit curve is shifted by 26.7 eV relative to the zero collision energy.

integrated parent ion yield at $U_{R2} = 2560$ V is relatively small compared to the increase of the fragment ion yields at the same surface potential (at $E_i = 160$ eV, see figure 4.63 b). Figure (4.63 b) shows the integrated fragment ion yields in dependence of the collision energy E_i of the three most intense fragments corresponding to $(CO)_{24}^+$ (○), $(CO)_{23}^+$ (◆) and $(CO)_{22}^+$ (◇). None of the integrated fragment ion yields show clear sigmoidal shape. The integrated fragment ion yield of the fragment corresponding to $(CO)_{24}^+$ shows a Gaussian shape and is well fitted by a Gaussian fit curve. The Gaussian fit curve yields a shift between the Gaussian peak maximum and zero collision energy of $\Delta E_i = 26.7$ eV (zero collision energy compared to the mean kinetic energy, see equation 4.3). Thus the Gaussian shape and the nearly symmetrical location of the integrated fragment ion yield around zero mean kinetic energy indicate impact induced fragmentation for the fragment corresponding to $(CO)_{24}^+$ (clear dependence on the mean kinetic energy distribution of the parent cluster). As mentioned before for all integrated fragment ion yields a jump in intensity is observed around $E_i = 160$ eV caused by an intensity jump in the integrated parent ion intensity. According to the mean kinetic energy of the parent ion ($E_0 = 2722$ eV) for the case of fragment generation by metastable decay in the field free region of the TOFMS the mean kinetic energy of the metastable fragments should be $E_d = 2613$ eV ($E_i = 109$ eV) for the $(CO)_{24}^+$, $E_d = 2504$ eV ($E_i = 218$ eV) for the $(CO)_{23}^+$ and $E_d = 2395$ eV ($E_i = 327$ eV) for the $(CO)_{22}^+$. However none of the fragments show indications for metastable decay in the field free region of the TOFMS even a clear sigmoidal shape. Only in the case of the fragment ion $(CO)_{24}^+$ a strong decrease around the value of $E_d = 2613$ eV ($E_i = 109$ eV) is observed. However the intensity of this fragment does not decrease to zero even at $E_i = 400$ eV (see also the next cluster size $(CO)_{26}^+$, figure 4.64). This behavior can be explained again by the superposition of the two different fragmentation effects, the metastable decay and the impact induced decay. In the case of the lighter two fragments the intensities do not change over a large energy range indicating metastable decay in the reflectron. The following cluster $(CO)_{26}^+$ behaves similar to the parent cluster ion $(CO)_{25}^+$. The stainless steel surface impact of the $(CO)_{26}^+$ parent cluster ion is depicted in the figure (4.64). The sigmoidal decrease of the integrated parent cluster ion yield is shown in the figure (4.64 a). The decrease of the parent ion yield is well fitted with an error fit function curve. Compared to the $(CO)_{25}^+$ cluster ion impact nearly the same mean kinetic energy is obtained by the fit curve ($E_0 = 2723.6$ eV). Here again an intensity shoulder around $U_{R2} = 2650$ V is present in the kinetic energy analysis graph. Three of the most intense fragment ion yields are shown in the figure (4.64 b) corresponding to the fragment ions $(CO)_{25}^+$, $(CO)_{24}^+$ and $(CO)_{23}^+$. The integrated fragment ion yield of the fragment ion corresponding to $(CO)_{25}^+$ shows a comparable shape to the fragment ion yield of the fragment corresponding to $(CO)_{24}^+$ in the figure (4.63 b). However, in the case of the integrated fragment ion yield of the fragment corresponding to $(CO)_{25}^+$ the sigmoidal decrease is more pronounced (notice the increase of the integrated fragment ion yield around the zero kinetic energy dashed line). Therefore a sigmoidal error function fit curve was used to determine the mean kinetic energy of that fragment. The fit curve yields a mean kinetic energy for the fragment corresponding to $(CO)_{25}^+$ of $E_d = 2631.7$ eV ($E_i = 104.8$ eV). According to this mean kinetic energy a mass difference of $\Delta m = 24.6$ amu is obtained between the parent cluster ion and the fragment cluster ion which should be 28 amu in the case of metastable decay in the field free region of the TOFMS. This discrepancy indicates a formation of these fragments in the reflectron collider. Another noticeable feature is that the intensity of the integrated fragment ion yield does not decrease to zero compared to the sigmoidal error function fit curve. Both observations indicate the superposition of two fragment formation effects, the fragment formation by metastable decay and the impact induced fragment formation. The integrated fragment ion yields of the two lighter fragments corresponding to $(CO)_{24}^+$ and

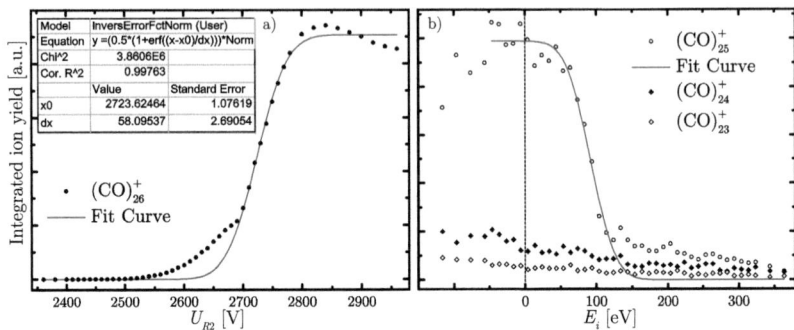

Figure 4.64 Collision of the $(CO)_{26}^+$ parent cluster ion with the stainless steel backplane of the reflectron collider. The reflectron TOFMS was operated in two stage mode at 3 kV extraction ($U_0 = 3$ kV, $U_1 = 2435$ V, $U_{R1} = 1647$ V and 300 eV EI with the valve mounted e-gun) for neat CO gas expansion ($P_0 = 3.5$ MPa and $T_0 = 305$ K). **a)** The decrease of the integrated ion yield with decreasing surface potential U_{R2} (•) with a sigmoidal error function fit curve (solid line). Here the reflectron collider is used as an energy analyzer. Similar to the $(CO)_{25}^+$ the decrease in the integrated ion yield shows one slight intensity shoulder around $U_{R2} = 2650$ V. **b)** Depicted are the integrated fragment ion yields of the fragments $(CO)_{25}^+$ (∘), $(CO)_{24}^+$ (♦) and the fragment $(CO)_{23}^+$ (◊) observed for the impact of the $(CO)_{26}^+$ parent cluster ion. Only the fragment ion yield of the heaviest fragment corresponding to $(CO)_{25}^+$ shows a decrease with nearly sigmoidal decrease with increasing collision energy E_i (see equation 4.3). The fragment ion yield of the fragment $(CO)_{25}^+$ shows a slight increase around the zero impact energy $E_i = 0$ eV (dashed line) which is an indication for impact induced fragmentation. The decrease of the integrated fragment ion yield of the fragment corresponding to $(CO)_{25}^+$ is well fitted by an error function fit curve except the decrease to zero ion yield. Contrary to the sigmoidal fit curve the integrated fragment ion yield does not decrease to zero which is an another indication for impact induced fragmentation.

$(CO)_{23}^+$ do not show sigmoidal shapes. Assuming metastable decay in the field free region the two lighter fragments would posses mean kinetic energies located around $E_d = 2514.1$ eV ($E_i = 209.5$ eV) in the case of $(CO)_{24}^+$ fragment and around $E_d = 2409.3$ eV ($E_i = 314.3$ eV) in the case of the $(CO)_{23}^+$ fragment. However no evidence for fragmentation in the field free region can be observed in the graph depicted in figure (4.64 b). Therefore it can be concluded that these lighter fragments are generated in the acceleration or deceleration region of the reflectron collider. The observed picture does not change dramatically in the case of the next larger cluster size, the $(CO)_{27}^+$ parent cluster ion. Figure (4.65) shows the results of the stainless steel surface impact of the $(CO)_{27}^+$ parent cluster ion. The decrease of the integrated parent cluster ion yield is similar to the measure-

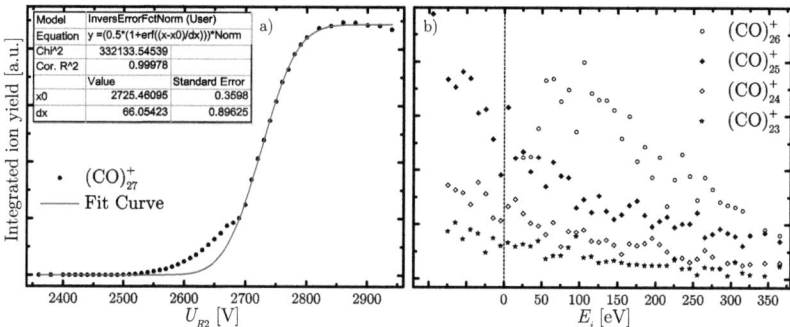

Figure 4.65 Collision of the $(CO)_{27}^+$ parent cluster ion with the stainless steel backplane of the reflectron collider. The reflectron TOFMS was operated in two stage mode at 3 kV extraction ($U_0 = 3$ kV, $U_1 = 2435$ V, $U_{R1} = 1647$ V and 300 eV EI with the valve mounted e-gun) for neat CO gas expansion ($P_0 = 3.5$ MPa and $T_0 = 305$ K). **a)** The decrease of the integrated ion yield with decreasing surface potential U_{R2} (•) with a sigmoidal error function fit curve (solid line). Here the reflectron collider is used as an energy analyzer. The decrease in the integrated ion yield shows one slight intensity shoulder (around $U_{R2} = 2650$ V) compared to the smaller size clusters with $n \leq 10$. **b)** Depicted are the integrated fragment ion yields of the fragment $(CO)_{26}^+$ (○), $(CO)_{25}^+$ (◆), $(CO)_{24}^+$ (◇) and the fragment $(CO)_{23}^+$ (★) observed for the impact of the $(CO)_{27}^+$ parent cluster ion. None of the fragment ion yields show clear sigmoidal behavior. The fragment ion yield of the fragment $(CO)_{26}^+$ increases with increasing impact energy up to the maximum located at $E_i \approx 110$ eV an indication for impact induced fragmentation. Additionally no fast decrease to zero integrated fragment ion yield was observed for all fragments. These behavior indicate the superposition of the two fragment generation effects (metastable decay and impact induced fragmentation) which was already observed in the case of the impact of the $(CO)_{25}^+$ parent cluster ion and in the case of the impact of the $(CO)_{26}^+$ parent cluster ion.

ments of the cluster sizes $(CO)_{25}^+$ and $(CO)_{26}^+$. Here again an intensity shoulder is observed around the surface potential of $U_{R2} = 2650$ V (see figure 4.65 a). Except the intensity shoulder the decrease in the integrated parent cluster ion yield is well fitted by an error function fit curve. The fit curve yields a mean kinetic energy of $E_0 = 2725.5$ eV which is in the same range of the mean kinetic energies obtained for $(CO)_{25}^+$ (see figure 4.63) and $(CO)_{26}^+$ (see figure 4.64). For stainless steel surface impact of the $(CO)_{27}^+$ parent cluster ion different fragment cluster ions were observed. Figure (4.65 b) shows the integrated fragment cluster ion yields of the most intense fragments the $(CO)_{26}^+$ (○), $(CO)_{25}^+$ (◆), $(CO)_{24}^+$ (◇) and $(CO)_{23}^+$ (★). Here again only in the case of the fragment corresponding to the decay $(CO)_{n-1}^+$ (fragment $(CO)_{26}^+$) a clear dependence between the parent cluster ion impact energy E_i and the integrated fragment ion yield can be observed. The integrated fragment ion yield of the fragment corresponding to $(CO)_{26}^+$ increases with increasing impact energy up to the maximum located at $E_i \approx 110$ eV. For parent cluster ion impact energies above $E_i \approx 110$ eV the integrated fragment ion yield decreases again. However the decrease of the integrated fragment ion yield of the fragment corresponding to $(CO)_{26}^+$ shows no clear sigmoidal shape. Hence no fast decrease to zero integrated fragment ion yield is observed. This behavior indicates again the superposition of the two fragment generation effects which was already observed in the case of the impact of the $(CO)_{25}^+$ parent cluster ion (for the fragment corresponding to $(CO)_{24}^+$) and in the case of the impact of the $(CO)_{26}^+$ parent cluster ion (for the fragment corresponding to $(CO)_{25}^+$). The integrated fragment ion yields of the three lighter fragments corresponding to $(CO)_{25}^+$, $(CO)_{24}^+$ and $(CO)_{23}^+$ do not possess sigmoidal shapes. Assuming metastable decay in the field free region the three lighter fragments would possess mean kinetic energies located around $E_d = 2523.6$ eV ($E_i = 201.9$ eV) in the case of $(CO)_{25}^+$ fragment, around $E_d = 2422.7$ eV ($E_i = 302.8$ eV) in the case of $(CO)_{24}^+$ fragment and around $E_d = 2321.7$ eV ($E_i = 403.8.3$ eV) in the case of the $(CO)_{23}^+$ fragment. However no evidence for fragmentation in the field free region can be observed in the graph depicted in figure (4.65 b). Therefore it can be concluded that these lighter fragments are generated in the acceleration or deceleration region of the reflectron collider. The largest cluster size which was impacted on the stainless steel surface was the carbon monoxide cluster ion consisting of thirty molecules ($(CO)_{30}^+$). Figure (4.66) shows the kinetic energy analysis measurement of the parent cluster ion (see figure (4.66) a) and the integrated ion yields of the fragment ions (see figure (4.66) b). The decrease of the integrated parent cluster ion yield is well fitted by an error function fit cure except a slight intensity shoulder. Kinetic energy analysis of the parent cluster ions yields a mean kinetic energy of $E_0 = 2743.5$ eV which is in the same energy range of the kinetic energies of the smaller clusters ($(CO)_{25}^+$, $(CO)_{26}^+$ and $(CO)_{27}^+$). Noticeable is the less pronounced intensity shoulder indicating lower fragmentation of the larger cluster ions in the acceleration region of the TOFMS accelerator (despite the larger flight time). Compared to the most intense $(CO)_{27}^+$ cluster ion peak intensity the integrated ion yield of the cluster ion $(CO)_{30}^+$ is half as high. Due to the overall lower count rate of the parent cluster ion the number and count rate of the fragment ions observed are too small. In the case of the $(CO)_{30}^+$ parent cluster ion only two intense fragment ions are observed, the fragment ion corresponding to $(CO)_{29}^+C$ and the fragment ion corresponding to $(CO)_{29}^+$. The integrated fragment ion yields of these fragment ions are shown in figure (4.66 b). Only the fragment corresponding to $(CO)_{29}^+C$ shows clear sigmoidal behavior. However the mean kinetic energy is located near the kinetic energy of the parent cluster ion (the zero collision energy line crosses the half hight of the sigmoidal decrease, see figure 4.66 b). According to the mean kinetic energy ratio between the parent cluster ions and the fragment $(CO)_{29}^+C$ cluster ions the fragment ions are not generated by

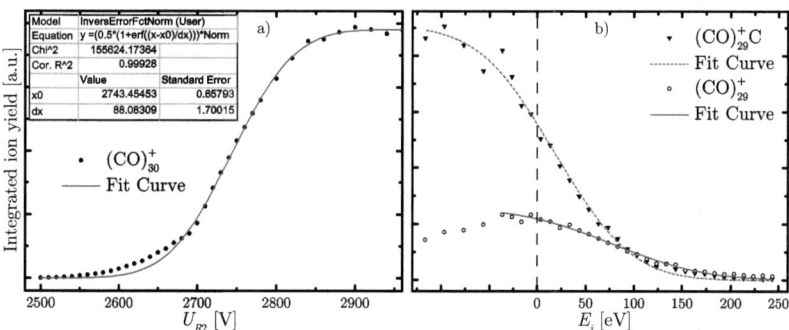

Figure 4.66 Collision of the $(CO)_{30}^+$ parent cluster ion with the stainless steel backplane of the reflectron collider. The reflectron TOFMS was operated in two stage mode at 3 kV extraction ($U_0 = 3$ kV, $U_1 = 2435$ V, $U_{R1} = 1647$ V and 300 eV EI with the valve mounted e-gun) for neat CO gas expansion ($P_0 = 3.5$ MPa and $T_0 = 305$ K). **a)** The decrease of the integrated ion yield with decreasing surface potential U_{R2} (•) with a sigmoidal error function fit curve (solid line). Here the reflectron collider is used as an energy analyzer. The decrease in the integrated ion yield shows one slight intensity shoulder (around $U_{R2} = 2650$ V) compared to the smaller size clusters with $n \leq 10$. **b)** Depicted are the integrated fragment ion yields of the fragment $(CO)_{29}^+C$ (▼) and the fragment $(CO)_{29}^+$ (○) observed for the impact of the $(CO)_{30}^+$ parent cluster ion. Only the fragment corresponding to $(CO)_{29}^+C$ shows sigmoidal behavior. The fragment ion $(CO)_{29}^+$ corresponding to monomer loss shows a intensity maximum located around zero impact energy $E_i = 0$ eV (see equation 4.3) an indication for impact induced fragmentation. The integrated ion yield of the fragment corresponding to $(CO)_{29}^+$ is well fitted by an error function except for impact energies $E_i \geq 160$ eV.

metastable fragmentation in the field free region of the TOFMS ($E_0 = 2743.5$ eV, $E_d = 2725.2$ eV ⟶ $\Delta m = 6$ amu). Hence these cluster ions must be already present in the acceleration region of the TOFMS accelerator during ion extraction. In the case of the lighter fragment ion corresponding to monomer loss the $(CO)_{29}^+$ cluster ion the integrated fragment ion yield shows a different behavior. The integrated fragment ion yield of the fragment corresponding to $(CO)_{29}^+$ shows a maximum value located around zero collision energy (dashed line at $E_i = 0$ eV see figure 4.66). The integrated fragment cluster ion yield of the fragment ion corresponding to $(CO)_{29}^+$ decreases with increasing collision energy E_i. The decrease of the integrated fragment ion yield has a sigmoidal shape for $E_i > 0$ eV and is well fitted by an error function fit curve. According to the mean kinetic energy ratio between the parent cluster ions and the fragment $(CO)_{29}^+$ cluster ions the fragment ions are not generated by metastable decay in the field free region of the TOFMS ($E_0 = 2743.5$ eV, $E_d = 2665.2$ eV ⟶ $\Delta m = 24$ amu). However a discrepancy is observed for higher collision energies ($E_i \geq 160$ eV) where the fit curve decreases fast to zero ion yield and the integrated fragment ion yield is quite higher than predicted by the fit curve. The maximal integrated fragment ion yield located around zero collision energy and the slight decrease to zero ion yield indicate again the superposition of the two fragment generation effects which were observed in the case of the other larger parent cluster ion sizes too ($(CO)_{25}^+$, $(CO)_{26}^+$ and $(CO)_{27}^+$).

4.3.3 Impact of $(CO)_n^+$ on SiO_2 covered $Si(100)$ Surface

In the last series of experiments the interaction of carbon monoxide cluster ions with a well defined silicon surface (see 4.1.5) was studied. In the previous subsection (see 4.3.2) carbon monoxide clusters were impacted on the stainless steel backplane of the reflectron collider. It was assumed that two different effects are involved in fragment generation, the metastable decay and surface impact induced decay. However it was not possible to distinguish between these two fragment types without doubts. Hence the experimental apparatus (as shown in figure 3.1) was modified with some new additions to distinguish between these two fragment types. The stainless backplane of the reflectron was removed and replaced by a wire mesh glued on the last ring electrode (which still defines the potential U_{R2}). Behind the mesh the new silicon surface was placed in a stainless steel disc surface holder (0.5 mm thickness, see subsection 4.1.5 and figure 4.32, based on the design in [285]). Compared to stainless steel the SiO_2 covered $Si(100)$ surface is well defined and possesses a high electron work function ϕ ($\phi_{Fe} = 4.5$ eV and $\phi_{Si(100)} = 4.9$ eV)[6]. Due to the higher work function ϕ of the SiO_2 surface the amount of impact neutralized cluster cations can be decreased increasing product yield. Therefore silicon surfaces were preferentially used in cluster surface scattering experiments [192; 201; 203–205; 220]. By this new configuration the silicon surface is located $d = 0.8$ mm behind the last reflectron mesh. A surface potential U_S can be applied to the silicon surface. To avoid charging of the silicon surface the surface is grounded over a 40 MΩ resistor chain. According to the new surface potential the collision energy E_i is defined by the difference between the mean kinetic energy E_0 (obtained by sigmoidal fit curves, see also figure 4.48) and the surface potential U_S. Hence equation (4.3) can be changed to the following form:

$$E_i = E_0 - eU_S. \tag{4.5}$$

Additional to the new surface and surface holder a new surface heater consisting of a commercial 150 W halogen lamp was placed behind the surface. With this heater the surface can be heated in a short time (several ten seconds) above 420 K to remove adsorbates from the surface. During the measurements the surface heater was operated in constant current mode to keep the desorption temperature around 420 K. The potential difference between the last reflectron electrode and the surface potential defines the extraction potential ($U_{Ex} = U_S - U_{R2}$) of surface impact products. The potential difference U_{Ex} between the surface potential and the last reflectron potential was optimized to obtain the highest fragment ion yield (100V $\leq U_{Ex} \leq$ 200 V). In that case the highest electric field strength which was necessary for impact induced product extraction was $U_S/d = 250$ V/mm. In the configuration used for the stainless steel surface impact the extraction field

---

[6]D. R. Linde, editor. *CRC Handbook of Chemistry and Physics*, CRC Press, Boca Raton, 76th. edition, 1995, 12-122/123.

strength changed with the U_{R2} value. The maximum potential value for U_{R2} was $U_{R2} = 3050$ V (in the case of 3 kV acceleration) and U_{R1} fixed at around $U_{R1} = 1650$ V. With these values the maximal value of the electric field strength was $(U_{R2} - U_{R1})/L_{R2} \approx 19$ V/mm. The extraction electric field strength of the later configuration with 250 V/mm is approximately by a factor of 12.5 larger than the maximal value of 19 V/mm of the former configuration. This high electric field strength value would improve the collection efficiency of the product ions formed after the impact. During the measurements the value of U_S and U_{R2} is lowered to increase the impact energy as known from the previous configuration. However the difference between these two voltages is kept constant to remain the same extraction potential U_{Ex}. Another difference to the previous configuration is the implementation of a rotatable energy analyzer in front of the detector (see figure A.5 in the appendix A.1.1). This energy analyzer[7] allows in beam retarding field energy analysis to distinguish between surface impact induced products ($E_d \approx eU_S$) and metastable decay products. Another advantage of this energy analyzer is the rotatable design which allows to remove the energy analyzer (with a magnetic rotary motion manipulator) from the ion beamline to increase transmission if energy analysis is not required or necessary.

In the following we will present some representative results of the surface impact of size selected carbon monoxide cluster ions on the SiO_2 passivated silicon surface. Small carbon monoxide cluster ions $(CO)_n^+$ up to the size $n = 15$ were generated with the flange mounted e-gun. Bigger carbon monoxide cluster ions $(CO)_n^+$ up to the size $n = 40$ were generated with the valve mounted e-gun. For similar expansion conditions the highest parent cluster ion signal intensity was observed for the carbon monoxide octamer $(CO)_8^+$ (with 6000 counts/min) and the lowest parent cluster ion signal intensity was observed for the carbon monoxide pentamer $(CO)_8^+$ (with 1270 counts/min) respectively. The most intense smallest cluster size was the pentamer. For the pentamer impact two groups of fragment ions were observed. One fragment ion peak has a narrow peak width (well resolved) whereas the other fragment ion peak has a very broad shape. Later energy analysis with the rotatable energy analyzer showed that the fragment ion peak with narrow peak width corresponds to metastable decay products $((CO)_4^+C$, see figure 4.67 b). Compared to these fragment ions the other fragment ions with the very broad peak shape posses a mean kinetic energy equal to the surface potential and could be identified as impact induced fragmentation products. This result is also confirmed by the behavior of the integrated fragment ion yield (see figure 4.67 b). Figure (4.67) depicts the surface impact results of the carbon monoxide pentamer cation on the SiO_2 covered silicon surface. In this case the parent cluster ions were generated at 150 V ionization potential with the flange mounted e-gun. Additionally the ionization parameters (e-gun parameters and e-gun to valve distance) were optimized to decrease fragmentation. Hence the decrease of

---

[7]Designed by Björn Kobin

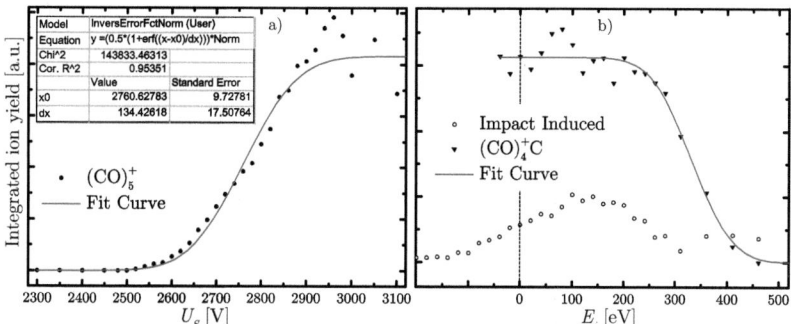

Figure 4.67 Collision of the $(CO)_5^+$ parent cluster ion with the SiO_2 covered silicon surface. The reflectron TOFMS was operated in two stage mode at 3 kV extraction ($U_0 = 3$ kV, $U_1 = 2435$ V, $U_{R1} = 1649$ V, $U_{Ex} = 100$ V and 150 eV EI with the flange mounted e-gun) for neat CO gas expansion ($P_0 = 3.0$ MPa and $T_0 = 333$ K, surface heater on). **a)** The decrease of the integrated ion yield of the parent cluster ion with decreasing surface potential U_S (●) with a sigmoidal error function fit curve (solid line). Here the reflectron collider was used as an energy analyzer. The decrease in the integrated ion yield shows intensity fluctuations and a comparable slight intensity shoulder. **b)** Depicted are the integrated fragment ion yields of the surface impact induced fragment ions (○, see text about products) and the metastable decay product $(CO)_4^+C$ (▼) observed for the impact of the $(CO)_5^+$ parent cluster ion. The fragment ion yield corresponding to metastable decay shows a sigmoidal decrease which is well fitted by an error function fit curve. Metastable decay of the $(CO)_{5+}$ parent cluster ion by O-atom loss is also confirmed by the mean kinetic energy obtained by the fit curve. The fragment ion yield of the impact induced fragments shows an increase with increasing impact energy E_i. Maximal ion yield is obtained for impact energies around $E_i = 100$ eV and decreases again. Noticeable is the second increase above $E_i = 300$ eV impact energy.

the integrated parent cluster ion yield is much smoother with less pronounced intensity shoulder than in the case of $(CO)_n^+$ cluster ions impacted on the stainless steel surface (see subsection 4.3.2). However some intensity fluctuations are observed for the high surface potential range with $U_S \geq 2900$ V. Note that here the second reflectron potential is given by $U_{R2} = U_S - 100$ V. Regarding the surface impact induced fragment ion masses, the peak width covers a large mass range. The mass range of the surface impact induced fragment ions begins with the mass corresponding to O-atom loss ($(CO)_4^+C$) and ends nearly at the parent cluster ion $(CO)_5^+$ peak. Additionally a slight peak is observed around the mass corresponding to C-atom loss ($(CO)_4^+O$). These observations support some new conclusions. One of the carbon monoxide molecules in the cluster is dissociated by the surface collision. The impact heated cluster cools down by the loss of a carbon-atom or oxygen-atom. However this process is slow enough to occur in the reflectron (delayed decay, not field free) which explains the large mass range and peak width observed for the surface impact induced fragments. The integrated fragment ion yield increases with increasing impact energy E_i. Maximum integrated fragment ion yield for the surface impact induced fragments is located around $E_i = 100$ eV. The amount of surface impact induced fragment ions decreases above $E_i = 200$ eV collision energy then again. For the impact induced fragment ion yield a second increase is observed above $E_i = 300$ eV. Compared to the behavior of the integrated fragment ion yields of the fragment ions corresponding to surface impact induced products the integrated fragment ion yield of the fragments corresponding to metastable decay products show clear sigmoidal behavior. The decrease of the integrated fragment ion yield of the fragment ions corresponding to metastable decay products is well fitted by an error function fit curve (see solid curve in figure 4.67 b). The fit curve yields a mean kinetic energy for the metastable decay products which correspond to O-atom loss $((CO)_4^+O)$. Another small peak corresponding to metastable monomer loss is observed, too (see figure 4.68). However, the count rate was too small for further analysis. In some cases peak analysis is complicated by the overlap of the fragment peaks with each other or the parent cluster ion peak (e. g. at higher surface potentials). In the case of the pentamer impact these peaks are more or less distinguishable for surface potentials below $U_S \leq 2800$ V. Figure 4.68 depicts the peak shape of the pentamer parent cluster ion and the peak shapes of the fragment ion peaks for two different collision energies ($E_i = 0$ eV for figure 4.68 a and $E_i = 100$ eV for figure b). The fragment ion peak corresponding to metastable O-atom loss overlaps with the parent cluster ion peak (figure 4.68 a) and shifts with increasing impact energy to lighter masses (this peak separates from the parent cluster ion peak). Fragment ion peaks corresponding to metastable decay are well resolved compared to the fragment ion peak related to surface impact induced fragment products. The peak corresponding to surface impact induced fragments covers a large mass range (123 amu – 137 amu, see figure 4.68). Retarding field energy analysis of the surface impact induced fragment peak showed that the fragment ions which cause the counts in the higher mass range posses less kinetic energy than the ions which appear in the lower mass range (see figure 4.69). This behavior indicates that the fragment ions which cause the counts in the higher mass range are delayed decay products which start later in the reflectron at a lower acceleration potential. Hence these ions appear in the higher mass range due to low kinetic energy and increased TOF to the detector. However this behavior complicates the interpretation of the observed surface impact induced fragment ions. The observed picture does not change in the case of the scattering of larger cluster ions e. g. the next cluster size the $(CO)_6^+$. Figure (4.70) depicts the surface impact results obtained for the impact of the hexamer $(CO)_6^+$ on the SiO_2 passivated silicon surface. The behaviors of the integrated fragment ion yields (see figure 4.70 a) are comparable to the behaviors of the integrated fragment ion yields of the

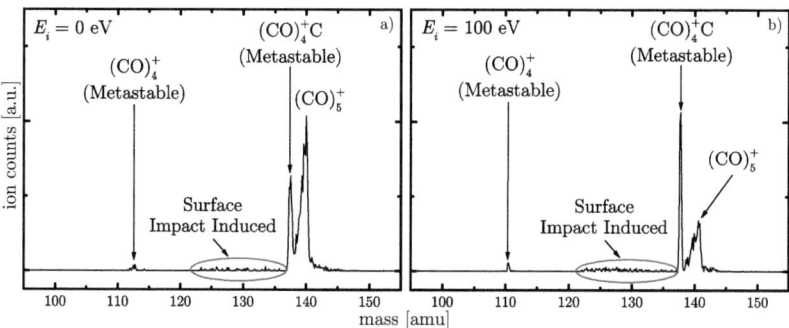

Figure 4.68 Mass spectra for two different collision energies of the $(CO)_5^+$ parent cluster ion with the SiO_2 covered silicon surface. The reflectron TOFMS was operated in two stage mode at 3 kV extraction ($U_0 = 3$ kV, $U_1 = 2435$ V, $U_{R1} = 1649$ V, $U_{Ex} = 100$ V and 150 eV EI with the flange mounted e-gun) for neat CO gas expansion ($P_0 = 3.0$ MPa and $T_0 = 333$ K, surface heater on). With increasing collision energy the parent cluster ion peak vanishes. Visible are the two peak shapes corresponding to the different fragmentation processes. Metastable fragment ions ($(CO)_4^+C$ and $(CO)_4^+$) posses peaks with narrow peak width which even get better resolved at higher collision energies. Fragmentation products corresponding to surface collision induced fragment ions are circled (○ in figure 4.67 b). Noticeable is the peak width which corresponds to a large mass range ($(CO)_4^+C - CO)_5^+$). **a)** Depicted are the peak shapes at $E_i = 0$ eV collision energy (related to the mean kinetic energy). **b)** Depicted are the peak shapes at $E_i = 100$ eV collision energy (related to the mean kinetic energy).

pentamer $(CO)_5^+$ silicon surface impact (see figure 4.67 b). For the surface impact induced fragment ions (○ in figure 4.70 a) an increasing integrated fragment ion yield is observed which is maximal around $E_i = 60$ eV. The integrated ion yield of the surface impact induced fragments decreases again above an impact energy of $E_i = 110$ eV. Noticeable is the second increase of the integrated fragment ion yield located around $E_i = 250$ eV of the surface impact induced fragment ions. A similar increase was observed in the case of the pentamer parent cluster ion silicon surface impact which was located around $E_i = 300$ eV collision energy (see figure 4.67 b). The integrated fragment ion yield of the other most intense fragment ion (triangles in figure 4.70 a) shows sigmoidal behavior. According to the mean kinetic energy obtained by an error function fit curve these fragment ions could be identified as $(CO)_5^+C$ metastable decay products which are formed by O-atom loss in the field free drift region of the TOFMS. The metastable decay origin of these fragments corresponding to O-atom loss is also confirmed by kinetic energy analysis with the rotatable retarding field energy analyzer in front of the MCP-detector (see figure 4.70 b). In figure (4.70 b) all potentials were hold at constant

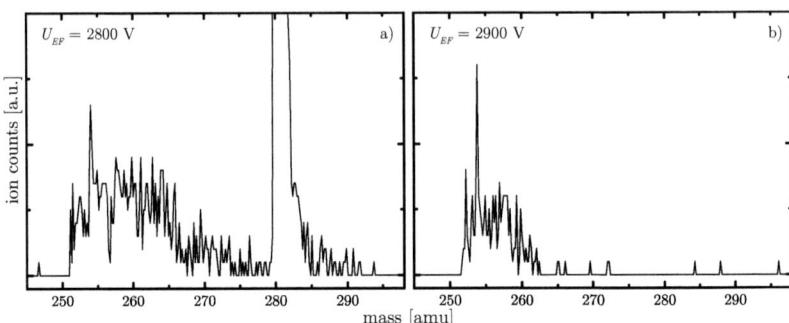

Figure 4.69 Mass spectra of the $(CO)_{10}^+$ parent cluster ion impact with the SiO_2 covered silicon surface for two different retarding field potentials U_{EF}. The reflectron TOFMS was operated in two stage mode at 3 kV extraction ($U_0 = 3$ kV, $U_1 = 2435$ V, $U_{R1} = 1649$ V, $U_{R2} = 2860$ V, $U_S = 2870$ V, $U_{Ex} = 10$ V and 150 eV EI with the flange mounted e-gun) for neat CO gas expansion ($P_0 = 3.5$ MPa and $T_0 = 303$ K, surface heater on). With increasing energy filter potential U_{EF} the parent cluster ion peak and the "heavier" surface impact induced fragmentation products disappear. **a)** Depicted is the surface impact mass spectrum for a energy filter potential $U_{EF} = 2800$ V (below the surface potential $U_S = 2870$ V). Besides the main parent $(CO)_{10}^+$ cluster peak the broad surface impact induced fragment peak is visible in the mass spectrum. **b)** Depicted is a mass spectrum for a retarding field potential $U_{EF} = 2900$ V, a little bit higher than the surface potential $U_S = 2870$ V. The main parent cluster peak disappears completely from the mass spectrum. Only a fraction of surface impact induced fragments corresponding to the lighter decay products are visible in the mass spectrum. The lighter impact induced fragment peak fraction begins at a mass value which corresponds to monomer loss ($m_{(CO)_9^+} = 252$ amu). According to this the heavier fraction of the surface impact induced fragment ions possesses less kinetic energy than the lighter fraction. Hence a delayed metastable decay of the surface impact heated parent ions in the reflectron cannot be excluded.

values ($U_{R1} = 1649$ V, $U_{R2} = 2600$ V and $U_S = 2700$ V) except the retarding potential of the energy analyzer (U_{EF}). Noticeable is the mean kinetic energy of the fragment ions corresponding to surface impact fragmentation products with $E_d = 2688.7$ eV (with a dx value of 72.5 eV) which is even higher (in this configuration) than the mean kinetic energy of the parent ions with $E_p = 2581.6$ (with a dx value of 77.5 eV). According to these mean kinetic energy values most of the parent cluster ions are reflected in the mesh region of the second reflectron stage with the potential value $U_{R2} = 2600$ eV. By this finding it can be concluded that the ions which form the main parent cluster peak (in this case the $(CO)_6^+$ parent cluster ions) originate from parent cluster ions which do not interact with

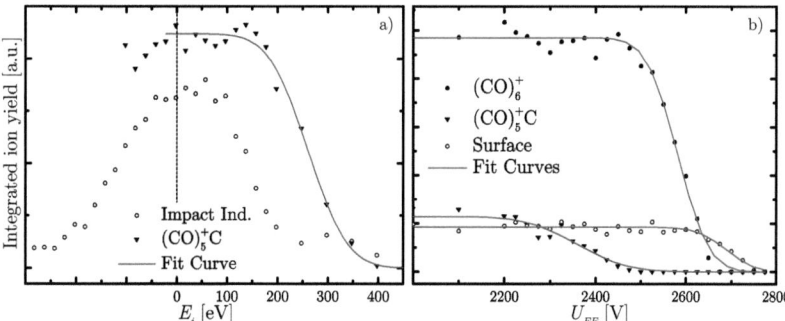

Figure 4.70 Collision of the $(CO)_6^+$ parent cluster ion with the SiO_2 covered silicon surface. The reflectron TOFMS was operated in two stage mode at 3 kV extraction ($U_0 = 3$ kV, $U_1 = 2435$ V, $U_{R1} = 1649$ V, $U_{Ex} = 100$ V and 150 eV EI with the flange mounted e-gun) for neat CO gas expansion ($P_0 = 3.5$ MPa and $T_0 = 313$ K, surface heater on). **a)** Depicted are the integrated fragment ion yields of the surface impact induced fragment ions (○) and the metastable decay product $(CO)_5^+C$ (▼) observed for the impact of the $(CO)_6^+$ (●) parent cluster ion. The fragment ion yield corresponding to metastable decay shows a sigmoidal decrease which is well fitted by an error function fit curve. Metastable decay of the $(CO)_6^+$ parent cluster ion by O-atom loss is also confirmed by the mean kinetic energies of the parent and fragment ions. The ion yield of the impact induced fragment ions shows a nearly Gaussian distribution located around zero impact energy $E_i = 0$ eV. Maximal ion yield is obtained for impact energies around $E_i = 60$ eV which decreases again above $E_i = 110$ eV. Noticeable is the second increase above $E_i = 250$ eV impact energy (see also figure 4.67). **b)** Kinetic energy analysis of the ion peaks for a given surface potential (U_S) with the rotatable retarding field energy filter (U_{EF}: energy filter potential) in front of the MCP-detector ($U_{R2} = 2600$ V, $U_S = 2700$ V). The origin of the ions can be distinguished by the mean kinetic energy of the ions obtained by sigmoidal fit curves. Three different error function fit curves yield three different mean kinetic energies for the peaks observed in the mass spectra (similar mass spectra as in figure 4.68). Interesting is the mean kinetic energy value obtained for the fragments assumed to be surface impact induced fragments (○). These fragment ions posses a mean kinetic energy value of $E_d = 2688.7$ eV (with a dx value of 72.5 eV) which matches well with the applied surface potential of $U_S = 2700$ V.

the silicon surface. In the other case the fragment ion peak which covers a broad mass range, corresponds to the surface impact generated fragmentation products with monomer loss $(CO)_{n-1}^+$, O-atom loss $(CO)_{n-1}^+C$, C-atom loss $(CO)_{n-1}^+O$ up to the intact scattered parent cluster ion $(CO)_n^+$ (in the case of smaller clusters, see also figure 4.69). With growing cluster size the mass range of the surface impact induced fragment ions increases too. In the case of the largest impacted parent cluster ions the $(CO)_{40}^+$ fragment masses with mass differences corresponding to pentamer loss were observed. However the loss of much larger fragments or a complete shattering of the parent cluster ion into smaller fragments e. g. monomers or dimers was not observed for all cluster sizes. The observed mass range of the surface impact induced fragment ions does not significantly change or shift with increasing collision energy. Hence only an increase in the intensity of the lighter fragment ion yield accompanied by a decrease or vanishing of the heavier fragment ion yield (peaks near the parent cluster ion peak) with increasing collision energy was observed. To confirm the different origin of the recorded peaks another set of kinetic energy analysis experiments was performed. In the following the scattering of the carbon monoxide octamer $(CO)_8^+$ ions with the SiO_2 covered silicon surface will be given as a representative result for these measurements. Here again the rotatable retarding field energy analyzer in front of the MCP-detector was utilized for mean kinetic energy determination. As in the case of the hexamer ion $(CO)_6^+$ (see figure 4.70 b) mean kinetic energy analysis all potentials were hold at constant values except the retarding potential of the energy analyzer (U_{EF}). However the main difference to the measurement with the hexamer ion was the utilization of two different potential configurations of the surface potential U_S and second reflectron stage potential U_{R2} (both with $\Delta U_{Ex} = 100$ V). Figure (4.71 a) depicts the kinetic energy analysis for the surface potential configuration with $U_S = 2640$ and second reflectron stage potential $U_{R2} = 2540$ V. The potentials were increased by 60 V to $U_S = 2700$ V and $U_{R2} = 2600$ V in the second configuration shown in figure (4.71 b). In figure (4.71) the mean kinetic energies of the three peaks corresponding to the parent ion, the metastable decay product and the surface impact products (broad mass peak) were analyzed for the two different potential configurations. The integrated ion yields of these peaks were fitted by error function fit curves. As it was expected the mean kinetic energy of the peak corresponding to monomer loss does not depend on the surface potential U_S and the second reflectron stage potential U_{R2}. For these fragment ions the fit curves yield a mean kinetic energy value around $E_d = 2338$ V ($E_d = 2337.7$ eV and dx $= 110.7$ eV in figure 4.71 a and $E_d = 2338.6$ eV with dx $= 117.4$ eV in figure 4.71 b). This is a clear proof for the metastable decay origin of these fragment ions which do not interact with the surface or second reflectron stage mesh. Regarding the parent ions, the assumptions made before for the hexamer are confirmed by the shift of the mean kinetic energy following the shift of the potential U_{R2}. For the two different configurations two different mean kinetic energy values were obtained for the parent ion

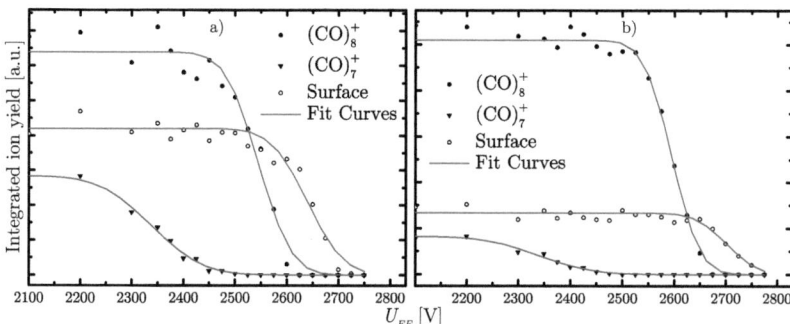

Figure 4.71 Collision of the $(CO)_8^+$ parent cluster ion with the SiO_2 covered silicon surface. The reflectron TOFMS was operated in two stage mode at 3 kV extraction ($U_0 = 3$ kV, $U_1 = 2435$ V, $U_{R1} = 1649$ V, $U_{Ex} = 100$ V and 150 eV EI with the flange mounted e-gun) for neat CO gas expansion ($P_0 = 3.5$ MPa and $T_0 = 313$ K, surface heater on). **a)** Kinetic energy analysis of the ion peaks (parent cluster ion $(CO)_8^+$ •, metastable decay product $(CO)_7^+$ ▼ and surface impact products ○) for a given surface potential $U_S = 2640$ V ($U_{R2} = 2540$ V), with the rotatable retarding field energy filter (U_{EF}). The origin of the ions can be distinguished by the mean kinetic energy of the ions obtained by sigmoidal fit curves. Three different error function fit curves yield three different mean kinetic energies (E) for the peaks observed in the mass spectra ($(CO)_8^+$: $E_p = 2545.4$ eV with dx = 71.5 eV, $(CO)_7^+$: $E_d = 2337.7$ eV with dx = 110.7 eV and surface products: $E_s = 2639.7$ eV with dx = 79.0 eV). Interesting is the mean kinetic energy value obtained for the fragments assumed to be surface impact induced fragments (○). These fragment ions posses a mean kinetic energy value of $E_s = 2639.7$ eV (dx = 79.0 eV) which matches well with the applied surface potential of $U_S = 2640$ V. **b)** Increasing the surface potential U_S and the second stage potential U_{R2} by 60 V ($U_S = 2700$ V and $U_{R2} = 2600$ V) the mean kinetic energy values increase too, except the mean kinetic energy value of the metastable decay product $(CO)_7^+$. The mean kinetic energy values shift to for $(CO)_8^+$: $E_p = 2595.4$ eV with dx = 61.0 eV and for the surface products: $E_s = 2701.3$ eV with dx = 66.2 eV. In contrast the mean kinetic energy value of the fragment $(CO)_7^+$ corresponding to metastable decay does not change and remains nearly at the same value: $E_d = 2338.6$ eV with dx = 117.4 eV.

peak ($U_{R2} = 2540$ V: $E_p = 2545.4$ eV with dx = 71.5 eV, in figure 4.71 a and $U_{R2} = 2600$ V: $E_p = 2595.4$ eV with dx = 61.0 eV, in figure 4.71 b). Similarities were observed for the mean kinetic energies obtained for the fragment ion peak corresponding to surface impact products. Hence in the case of the broad fragment ion peak the mean kinetic energy of the particles depends on the surface potential and shifts with U_S ($U_S = 2640$ V: $E_s = 2639.7$ eV with dx = 79.0 eV, in figure 4.71 a and $U_S = 2700$ V: $E_s = 2701.3$ eV with dx = 66.2 eV, in figure

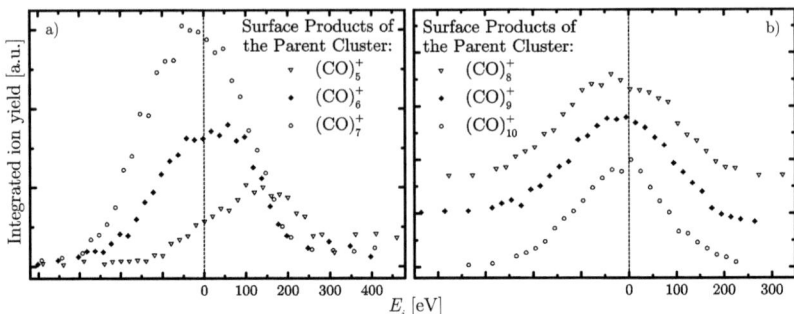

Figure 4.72 Integrated ion yields of the impact induced products for the surface collision of the $(CO)_5^+$–$(CO)_{10}^+$ parent cluster ions with the SiO_2 covered silicon surface. The reflectron TOFMS was operated in two stage mode at 3 kV extraction ($U_0 = 3$ kV, $U_1 = 2435$ V, $U_{R1} = 1649$ V, $U_{Ex} = 100$ V and 150 eV EI with the flange mounted e-gun) for neat CO gas expansion ($P_0 = 3.5$ MPa and $T_0 = 313$ K, surface heater on). **a)** Depicted are the integrated ion yields of the surface impact induced fragments ions recorded for three different parent cluster sizes: $(CO)_5^+$ ($\triangledown$), $(CO)_6^+$ ($\blacklozenge$) $(CO)_7^+$ ($\circ$). All yield curves are shown here with their "real" amplitude (not normalized). The maximum of the integrated product ion yields shifts with increasing cluster size to the zero collision energy line. **b)** Depicted are the integrated ion yields of the surface impact induced fragments ions recorded for three different parent cluster sizes: $(CO)_8^+$ ($\triangledown$), $(CO)_9^+$ ($\blacklozenge$) $(CO)_{10}^+$ ($\circ$). All yield curves are normalized to their maximum value and shifted by a constant y-value to form a stack graph. The curves are more or less centered around the zero impact energy line $E_i = 0$ eV. However, with increasing cluster size an increasing asymmetry of the integrated product ion yields is observed. The decrease of the integrated surface product ion yields with increasing impact energy E_i is steeper with increasing cluster size (see also figure 4.73).

4.71 b). These results confirm the different origins of the observed peaks and demonstrate the possibility to distinguish between metastable decay and surface impact induced fragmentation products. Due to the detailed discussion of the metastable decay products in the previous subsections (see 4.3.1 and 4.3.2), in the following we will limit our analysis and findings to the surface impact induced fragmentation products. Figure (4.72) depicts the surface impact results for the $(CO)_5^+$–$(CO)_{10}^+$ parent cluster ions. Displayed are the integrated ion yields of the surface impact induced product ions for the different cluster sizes in dependence of the collision energy E_i. In comparison to figure (4.72 b) in figure (4.72 a) the curves are shown with their "real" amplitude (not normalized). All integrated ion yield curves of the surface impact induced products shown in figure (4.72) have nearly Gaussian shapes. The most dramatic change of the integrated ion yield curves of the surface impact induced products are observed for the smaller parent cluster sizes shown in figure (4.72 a). In the case of the pentamer (the smallest sample parent cluster) ion the maximum of the integrated product ion yield is located around $E_i = 100$ eV. With increasing parent cluster ion size the curve maximum shifts to the zero collision energy line with $E_i = 0$ eV. This behavior confirms the higher stability of smaller carbon monoxide cluster ions. According to Mähnert et al. [13] the carbon monoxide dimer shows with 1.80 eV one of the largest reported binding energy values for an ionized van der Waals dimer (see also ref. [14] for small $(CO)_n^+$ clusters). With this shift of the maximum product ion yield value the largest E_i value at which surface products are barely observed decreases too. In the case of the heptamer $(CO)_7^+$ this E_i value is located around 400 eV (see figure 4.72 a). Compared to the heptamer in figure (4.72 a) the integrated product ion yield of the octamer is barely detectable around $E_i = 300$ eV (see figure 4.72 b). Note that in figure (4.72 b) the curves are normalized relative to their maximal value and shifted by a constant y-value to form a stack graph. All curves shown in figure (4.72 b) are more or less centered around the zero impact energy line $E_i = 0$ eV (± 20 eV). Hence the maximum values of the integrated product ion yields are located near the zero impact energy line. However with increasing cluster size an increasing asymmetry of the integrated product ion yields is observed. The decrease of the integrated surface product ion yields with increasing impact energy E_i is steeper with increasing cluster size (see also figure 4.73). Therefore the E_i value at which surface impact products are barely observed decreases further with increasing parent cluster size. In the case of the largest parent cluster ion in figure (4.72), the decamer $(CO)_{10}^+$, this value is located around $E = 240$ eV which is roughly half as high as the value observed for the pentamer $(CO)_5^+$. This behavior can be explained by the steep decrease of the binding energy with increasing cluster size as reported by Hiraoka et al. [14]. Thus it can be assumed that after a certain impact energy value E_i/n the parent cluster ion does not survive the impact and shatters [197; 208]. However as mentioned before no clear indication for shattering of the impinging parent cluster ions were observed in the mass spectra (no monomer or dimer fragment ions were observed, even for E_i values > 500 eV). Nevertheless it cannot be excluded that shattering occurs and escapes detection due to the experimental conditions (angular distribution, to my knowledge shattering was observed only with non common reflectron collider configurations e. g. [198; 208; 223; 286]). Similar results were obtained for the impact of larger clusters as shown in figure (4.73). However, regarding the larger clusters displayed in figure (4.73), the decrease of the impact energy value E_i where no or barely impact induced product ions are observed is much smaller compared to the decrease observed for the smaller cluster sizes (see figure 4.72). In the case of the $(CO)_{15}^+$ parent cluster ion the ion yield of the surface impact induced ions decreases to zero around $E_i = 200$ eV (see figure 4.73 b) which is not much lower than the value $E_i = 240$ eV observed for the decamer $(CO)_{10}^+$ (see figure 4.72 b). This value does not significantly change further with

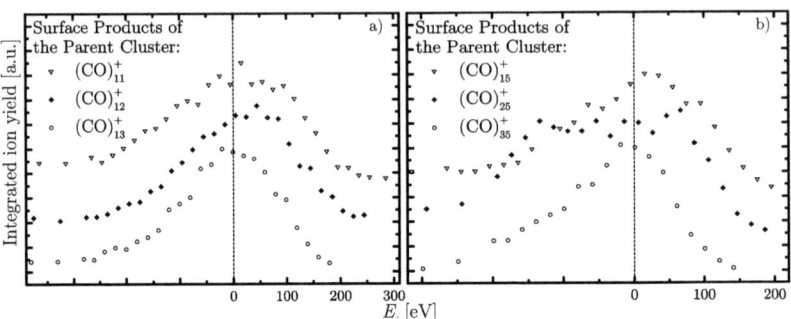

Figure 4.73 Integrated ion yields of the impact induced products for the surface collision of the $(CO)_{11}^+$–$(CO)_{35}^+$ parent cluster ions with the SiO_2 covered silicon surface. The reflectron TOFMS was operated in two stage mode at 3 kV extraction ($U_0 = 3$ kV, $U_1 = 2435$ V, $U_{R1} = 1649$ V, $U_{Ex} = 100$ V and 150 eV EI with the flange mounted e-gun, except the $(CO)_{25}^+$ and $(CO)_{35}^+$ cluster ions were generated with the valve mounted e-gun at 250 eV EI) for neat CO gas expansion ($P_0 = 3.5$ MPa, $P_0 = 2.5$ MPa in the case of $(CO)_{25}^+$ and $(CO)_{35}^+$, $T_0 = 313$ K, surface heater on). **a)** Depicted are the integrated ion yields of the surface impact induced fragments ions recorded for three different parent cluster sizes: $(CO)_{11}^+$ ($\triangledown$), $(CO)_{12}^+$ ($\blacklozenge$) and $(CO)_{13}^+$ ($\circ$). All yield curves are normalized to their maximal value and shifted by a constant y-value to form a stack graph. **b)** Depicted are the integrated ion yields of the surface impact induced fragment ions recorded for three different parent cluster sizes: $(CO)_{15}^+$ ($\triangledown$), $(CO)_{25}^+$ ($\blacklozenge$) and $(CO)_{35}^+$ ($\circ$). All yield curves are normalized to their maximal value and shifted by a constant y-value to form a stack graph. The curves are more or less centered around the zero impact energy line $E_i = 0$ eV. The decrease of the integrated surface product ion yields with increasing impact energy E_i is steeper with increasing cluster size (see also figure 4.72).

increasing cluster size up to the $(CO)_{25}^+$. A second small decrease is observed for the largest cluster size shown in figure (4.73 b), the $(CO)_{35}^+$. For the impact of the $(CO)_{35}^+$ parent cluster ion the integrated ion yield of the surface impact induced products decreases steeply to zero with increasing impact energy E_i. In the case of the $(CO)_{35}^+$ parent cluster ion the integrated ion yield of the surface impact products vanishes around an impact energy value of $E_i = 150$ eV. These threshold energy values E_i in dependence of the parent cluster size from where on no or barely impact induced product ions are observed are summarized in figure 4.74. Interesting is the behavior observed in the figure (4.74 b) which shows the same threshold energy values as in figure (4.74 a) divided by the parent cluster size (threshold energy per molecule). These values show a rapid decrease with increasing cluster size. Hence the data shown in figure (4.74 b) is well fitted by an exponential fit curve (solid line). The surface impact results confirm the higher

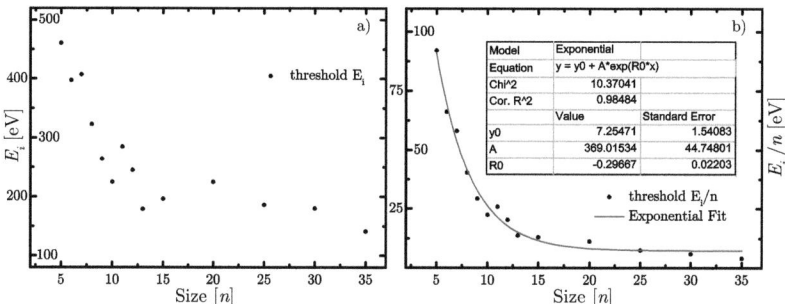

Figure 4.74 Depicted are cluster size dependent threshold impact energies E_i for the disappearance of the impact induced fragmentation ion yields for the surface collision of the $(CO)_5^+$–$(CO)_{35}^+$ parent cluster ions with the SiO_2 covered silicon surface. The threshold energies were extracted from the graphs shown before (see figure 4.72 and 4.73). Note that the mean kinetic energy distribution of all parent cluster ions is about $\Delta E_{\text{FWHM}} = 73.7$ eV (determined by Gaussian fit curves, average value with a standard deviation of $\sigma_{\Delta E} = 11.1$ eV). **a)** Depicted are the threshold energy values for the disappearance of the impact induced integrated fragment ion yields in dependence of the parent cluster size. **b)** Depicted are the threshold energy values divided by the cluster size n for the disappearance of the impact induced integrated fragment ion yields in dependence of the parent cluster size. The data is well fitted with an exponential fit curve (solid line).

stability of small carbon monoxide cluster ions $(CO)_n^+$ as reported in [13; 14]. The behavior of the graph shown in figure (4.74 b) is comparable with the decrease of the binding energy of $(CO)_n^+$ with increasing cluster size n as reported in ref. [14] for $(CO)_n^+$ with $3 \leq n \leq 18$.

Chapter 5
Summary and Outlook

> *"Finally, two days ago, I succeeded - not on account of my hard efforts, but by the grace of the Lord. Like a sudden flash of lightning, the riddle was solved. I am unable to say what was the conducting thread that connected what I previously knew with what made my success possible."* CARL FRIEDRICH GAUSS (1777–1855)

5.1 Summary

Re-TOFMS A compact, high resolution Re-TOFMS was designed, simulated, constructed and successfully tested. With the newly developed instrument molecular clusters were ionized, size selected and impacted on a well defined surface as a function of the collision energy. Two different electron guns were optimized for the generation of a broad sample cluster ion size range with $1 \leq n \leq 300$. The new Re-TOFMS apparatus is capable to mass select cluster ions up to the size $n = 190$. With the developed compact instrument mass spectra can be recorded with high resolution ($m/\Delta m > 3000$) even at moderate acceleration voltages of 4 kV and short total flight lengths of ≈ 1.5 m. The optimization of the electrode shapes allowed to construct a three stage TOFMS accelerator consisting of only six electrodes and three meshes. Additionally the design of the TOFMS reflectron was simplified by the invention of an alternating electrode thickness configuration (a thicker electrode followed by a thinner electrode). This allows to construct ion optical devices with long homogeneous electric fields with less number of electrodes (nearly half the number of electrodes that would be required for an equivalent conventional stacked ring electrode configuration).

The reflectron collider with the stainless steel backplane was operated as an ion kinetic energy analyzer. Hence the origins of fragmentation products were determined by kinetic energy analysis. Later the stainless steel backplane was replaced by a mesh and a surface holder with a SiO_2 covered $Si(100)$ surface. Additionally a new rotatable kinetic energy analyzer was mounted in front of the MCP detector. With these modifications it was possible to distinguish between surface impact induced fragments and fragments generated by metastable decay of the hot clusters (induced by electron impact ionization and skimmer interference).

Impact of $(CO_2)_n^+$ on stainless steel For the stainless steel surface impact of small carbon dioxide cluster ions (up to the size $(CO_2)_{25}^+$) from the hyperthermal energy range and above (up to several hundred eV collision energy) no evidence for collision induced fragmentation, shattering and dissociation could be detected. A possible explanation for the lack of shattering could be the fact that small fragments e. g. monomers formed after the impact glide lateral to the surface [287] which would limit the detection of these fragments with the present reflectron collider setup. An additional explanation could be the low work function of clean steel surfaces and thus the high efficiency for neutralization of the colliding cations. In literature for a clean stainless steel surface a ten times lower surface impact induced (SID) fragment ion yield compared to an organic material covered surface was reported [281]. Hence, it was observed that most of the parent cluster ions were neutralized upon surface impact. Thus, the reflectron collider with the stainless steel backplane was used for kinetic energy analysis of the impinging cluster ions. In that case clear sigmoidal decrease of the integrated ion yield of the parent clusters was observed for all cluster sizes. Deviations from the sigmoidal shape in terms of intensity plateaus were interpreted as metastable fragmentation. Depending on the mean kinetic energy obtained from fits of the intensity plateaus (also with sigmoidal shape) it was possible to distinguish between fragments formed in the acceleration region (region 3 in figure 4.49) and fragments formed by metastable decay in the field free region (region 5 in figure 4.49). For all cluster sizes except the monomer parent cluster ion EI induced fragment ions with a mass corresponding to $(CO_2)_{n-1}^+ O_2$ were detected. Up to the hexamer these fragments show no clear dependence on the surface potential U_{R2} nor a sigmoidal decrease indicating formation in the acceleration region by metastable decay e. g. the metastable decay of a bigger cluster of the same series: $(CO_2)_n^+ O_2 \longrightarrow (CO_2)_{n-1}^+ O_2 + CO_2$. However, for cluster sizes larger than the hexamer e. g. the heptamer or octamer the integrated ion yields of these clusters show a sigmoidal decrease. According to the mean kinetic energy obtained by error fit functions these fragments are already present in the extracted beam as reported by other groups e. g. in [278] (indicating formation by EI induced dissociation in region 1 in figure 4.49). Another already well known fragmentation channel (reported for collision induced and electron impact induced fragmentation [163; 288]) corresponding to the loss of monomers $(CO_2)_n^+ \longrightarrow (CO_2)_{n-1}^+ + CO_2$ was observed for various cluster sizes beginning from the trimer. Surprisingly for cluster sizes larger than the hexamer a previously unknown fragmentation channel could be observed. According to the kinetic energy analysis clusters of the series $(CO_2)_n^+ \longrightarrow (CO_2)_{n-1}^+ CO + O$ seem to be formed by metastable decay in the field free region (pre-dissociated by EI [279] in region 1 in figure 4.49, formed by metastable decay in region 5 in figure 4.49). The intensity plateaus observed for the integrated ion yield of the parent clusters correspond to a mass difference of 16 amu (distinct in the case of the heptamer, octamer, nonamer and decamer). These intensity plateaus were observed for cluster sizes larger than the hexamer.

With increasing cluster size these plateaus are less pronounced and merge with the main sigmoidal decrease of the integrated parent ion yield. This behavior can be explained by the decreasing mean kinetic energy ratio between the parent and fragment daughter ion with increasing cluster size. According to these results carbon dioxide molecules are "pre-dissociated" by EI and the fragments remain in the "hot" metastable clusters which may decay in the field free region.

Impact of $(CO)_n^+$ on stainless steel In two series of measurements small carbon monoxide cluster ions $(CO)_n^+$ with $2 \leq n \leq 10$ and with $25 \leq n \leq 30$ were impacted on the stainless steel surface backplane of the reflectron collider. The smaller cluster ions with $2 \leq n \leq 10$ were generated with the flange mounted e-gun. The larger cluster ions with $25 \leq n \leq 30$ were generated with the valve mounted e-gun. Noticeable no carbon monoxide monomer ion peaks were observed in the mass spectra. In the case of ion generation with the flange mounted e-gun heavy fragmentation in the TOFMS accelerator was observed. The absence of the monomer and the heavy fragmentation in the TOFMS accelerator indicate that small clusters are formed by fragmentation of larger clusters via EI. Therefore, these clusters are "hot" and cool down by successive evaporation of x monomers $(CO)_n^+ \longrightarrow (CO)_{n-x}^+ + x \cdot CO$. Hence, for the smaller parent clusters $n \leq 5$ heavy fragmentation in the acceleration region of the TOFMS was observed. Depending on the experimental observation time window metastable decay with successive monomer loss $(CO)_n^+ \longrightarrow (CO)_{n-x}^+ + x \cdot CO$ up to $x = 5$ monomers was observed for the small clusters $2 \leq n \leq 10$ (x increased with the parent cluster size). The fragment peak corresponding to one monomer loss ($x = 1$) could be identified as a metastable decay product formed in the field free region of the TOFMS (region 5 in figure 4.49). According to the mean kinetic energy of the fragment peaks corresponding to the loss of several ($x \geq 2$) monomers these fragments are formed by metastable decay in the reflectron (during deceleration or acceleration). In that sense the heavy metastable decay of carbon monoxide cluster ions could be a possible reason that this molecule was not studied till now in cluster-surface impact experiments.

Similar behavior was observed for the fragments corresponding to the mass of $(CO)_{n-1}^+ C$. As in the case of $(CO_2)_{n-1}^+ O_2$ these fragments seem to be EI pre-dissociated products which decay in the TOFMS accelerator or the reflectron. A clear indication for the formation of surface impact induced fragments or shattering was not observed for the smaller cluster sizes $2 \leq n \leq 10$ (similar explanation as in the case of $(CO_2)_n^+$ impacted on stainless steel surface). The situation changed by the use of the valve mounted e-gun. With the valve mounted e-gun $(CO)_n^+$ clusters with $25 \leq n \leq 30$ were generated. In the case of these clusters the heaviest fragment corresponding to monomer loss $(CO)_{n-1}^+$ showed indications for surface impact induced fragmentation. For these fragments the ion yield shows clear dependence on the impact energy E_i and the intensity maximum is located around zero impact energy ($E_i = 0$ eV, an indication for weakly bound clusters). Additionally, despite the fact that the intensity of these fragments shows a sigmoidal decrease the value does not decrease to zero fragment ion yield for higher impact energies (as expected for a sigmoidal decrease). As in the case of the smaller parent cluster ions no indication for the occurrence of a shattering event was observed. Besides, with the utilized experimental configuration it was not possible to clearly distinguish between the impact induced and EI induced fragmentation products. A solution for this problem was found and successfully applied to the surface impact of $(CO)_n^+$ on SiO_2 covered Si(100) surface (see next paragraph).

Impact of $(CO)_n^+$ on SiO_2 covered Si(100) Surface In the last series of experiments small carbon monoxide cluster ions $(CO)_n^+$ with $5 \leq n \leq 40$ were impacted on the SiO_2 covered Si(100) surface. Small carbon monoxide cluster ions $(CO)_n^+$ up to the size $n = 15$ were generated with the flange mounted e-gun. Bigger carbon monoxide cluster ions $(CO)_n^+$ up to the size $n = 40$ were generated with the valve mounted e-gun. Additionally the ionization parameters (e-gun parameters and e-gun to valve distance) were optimized to decrease EI induced fragmentation. However, it was not possible to totally eliminate EI induced fragmentation without significant parent cluster ion intensity loss. Hence a new different experimental configuration (based on the work [285]) was utilized to distinguish between surface impact induced fragmentation products and EI induced fragmentation products. A new surface holder for the silicon surface with a custom-made surface heater was implemented to the reflectron collider. Besides this a rotatable retarding field energy analyzer was mounted in front of the MCP detector. These modifications of the experimental setup enabled to set separate potentials for the surface (U_S) and for the second reflectron stage (U_{R2}). Hence these modifications of the experimental setup allowed distinguishing between metastable decay products (EI induced) and surface impact induced fragmentation products by kinetic energy analysis. As a result of these modifications a new peak with a broad mass range could be identified as surface impact induced fragmentation products. Fragment ion peaks corresponding to metastable decay are well resolved compared to the fragment ion peaks related to surface impact induced fragmentation products. Kinetic energy analysis of the surface impact induced fragment peaks showed that the lighter fraction of these peaks are fragmentation products formed near the surface (nearly surface potential U_S). Compared to these ions the ions present in the "heavier" part of the peak possess lower kinetic energies indicating formation during acceleration in the reflectron (delayed decay products). The lighter fraction of these impact induced fragmentation product peaks correspond to loss of monomers: $(CO)_n^+ \longrightarrow (CO)_{n-x}^+ + x \cdot CO$ with $1 \leq x \leq 5$ (the value of x increases with increasing parent cluster size, e. g. $x = 5$ was observed for the biggest impacted cluster with $n = 40$). For the surface impact induced fragmentation products the integrated fragment ion yield maximum shows a clear cluster size dependent shift to zero collision energy for the small clusters ($5 \leq n \leq 7$). As in the case of the other series of experiments for the scattering of carbon monoxide clusters on Si(100) no evidence for shattering transition was observed.

Interesting is also the shift of the threshold impact energy at which the surface impact induced fragmentation product ion yield vanishes. However, it is not clear whether this is a shattering transition and the shattering products are not detectable with the present reflectron collider setup. This threshold energy value shows clear parent cluster size dependent behavior and decreases nearly exponential with increasing size of the impacting parent cluster ion. As known from previous works [13; 14] this observation can be related to the steep decrease of the binding energy.

5.2 Outlook

Optimization of the TOFMS for 6 kV and three stage operation would further increase the resolving power of the present instrument. First attempts for 6 kV and three stage operation were successful. However, no attempts were made to further increase resolution by optimizing 6 kV three stage operation due to focusing on the scattering experiments. Resolution of the impact induced fragmentation products can be increased by decoupling the reflectron and the surface holder and scattering surface. The implementation of an additional mesh in front of the surface would allow to operate the surface collider as a two stage accelerator (pulsed operation, as described in [285]). This modification would allow the determination of the impact time of the parent cluster ion which would simplify the mass calibration for the scattering products. Expanding the apparatus with an additional manipulator for the scattering surface angle could increase scattering product yields. Additionally it would be possible to analyze scattering angle dependent phenomena. Due to the importance of the kinetic energy it would be desirable to decrease the kinetic energy distribution value to a minimum. However, this would require additional effort e.g. the utilization of LASER ionization or focusing the molecular beam with an einzel lens into the TOFMS accelerator. Another possibility would be to use a skimmer with a smaller diameter to reduce the spatial distribution of the ions in the accelerator and thus the kinetic energy distribution. However all of these methods have their drawbacks and limitations. A useful instrumental upgrade would be the implementation of a collision cell to compare collision induced dissociation (CID) with surface impact induced dissociation (SID). New promising model systems for experimental studies would be the surface scattering of mixed clusters. Mixed van der Waals bound clusters can be generated by seeded molecular beam expansion or by molecule pickup from an effusive beam. Accordingly, cluster impact induced intracluster chemical reactions can be studied for various combinations of molecules and surfaces e. g. catalytically active surfaces.

Bibliography

[1] D. Herschbach. Chemical physics: Molecular clouds, clusters, and corrals. *Rev. Mod. Phys.*, 71(2):S411–S418, 1999. doi: 10.1103/RevModPhys.71.S411.

[2] A. Largo, P. Redondo, and C. Barrientos. Theoretical study of possible ion-molecule reactions leading to precursors of glycine in the interstellar medium. *Int. J. Quantum Chem.*, 98(4):355–360, 2004. doi: 10.1002/qua.20070.

[3] M. Kulmala, I. Riipinen, M. Sipilä, H. E. Manninen, T. Petäjä, H. Junninen, M. Dal Maso, G. Mordas, A. Mirme, M. Vana, A. Hirsikko, L. Laakso, R. M. Harrison, I. Hanson, C. Leung, K. E. J. Lehtinen, and V.-M. Kerminen. Toward direct measurement of atmospheric nucleation. *Science*, 318(5847):89–92, 2007. doi: 10.1126/science.1144124.

[4] A. J. Cox, J. G. Louderback, and L. A. Bloomfield. Experimental observation of magnetism in rhodium clusters. *Phys. Rev. Lett.*, 71(6):923, 1993. doi: 10.1103/PhysRevLett.71.923.

[5] U. Heiz and W.-D. Schneider. Nanoassembled model catalysts. *J. Phys. D*, 33 (11):R85–R102, 2000. doi: 10.1088/0022-3727/33/11/201/.

[6] J. Bansmann, S. H. Baker, C. Binns, J. A. Blackman, J.-P. Bucher, J. Dorantes-Dávila, V. Dupuis, L. Favre, D. Kechrakos, A. Kleibert, K.-H. Meiwes-Broer, G. M. Pastor, A. Perez, O. Toulemonde, K. N. Trohidou, J. Tuaillon, and Y. Xie. Magnetic and structural properties of isolated and assembled clusters. *Surf. Sci. Rep.*, 56(6-7):189–275, 2005. doi: 10.1016/j.surfrep.2004.10.001.

[7] J. P. Wilcoxon and B. L. Abrams. Synthesis, structure and properties of metal nanoclusters. *Chem. Soc. Rev.*, 35(11):1162–1194, 2006. doi: 10.1039/b517312b.

[8] T. M. Bernhardt, U. Heiz, and U. Landman. *Nanocatalysis*, chapter Chemical and catalytic properties of size-selected free and supported clusters, pages 1–177. NanoScience and Technology. Springer, Berlin, 2007. ISBN 9783540326458. U. Heiz, U. Landman, (editors).

[9] A. W. Castleman, Jr. and S. N. Khanna. Clusters, superatoms, and building blocks of new materials. *J. Phys. Chem. C*, 113(7):2664–2675, 2009. doi: 10.1021/jp806850h.

[10] R. Moro, R. Rabinovitch, C. Xia, and V. V. Kresin. Electric dipole moments of water clusters from a beam deflection measurement. *Phys. Rev. Lett.*, 97(12): 123401–4, 2006. doi: 10.1103/PhysRevLett.97.123401.

Bibliography

[11] O. Ingolfsson and A. M. Wodtke. Electron attachment time-of-flight mass spectrometry reveals geometrical shell closings in van der Waals aggregates. *J. Chem. Phys.*, 117(8):3721–3732, 2002. doi: 10.1063/1.1495402.

[12] W. Kedzierski and J. W. McConkey. Fluorescence following electron impact on argon clusters. *J. Chem. Phys.*, 107(17):6521–6525, 1997. doi: 10.1063/1.474895.

[13] J. Mähnert, H. Baumgärtel, and K.-M. Weitzel. The investigation of the $(CO)_2^+$ ion by dissociative ionization of argon/carbon monoxide clusters. *J. Chem. Phys.*, 103(16):7016–7024, 1995. doi: 10.1063/1.470328.

[14] K. Hiraoka, T. Mori, and S. Yamabe. On the formation of the isomeric cluster ions $(CO)_n^+$. *J. Chem. Phys.*, 94(4):2697–2703, 1991. doi: 10.1063/1.459844.

[15] T. D. Märk. Cluster ions: Production, detection and stability. *Int. J. Mass Spectrom. Ion Processes*, 79(1):1–59, 1987. doi: 10.1016/0168-1176(87)80022-8.

[16] C. Yeretzian, R. D. Beck, and R. L. Whetten. Cluster-surface scattering in a reflectron collider: probing fullerenes by surface impact. *Int. J. Mass Spectrom. Ion Processes*, 135(2-3):79–118, 1994. doi: 10.1016/0168-1176(94)04011-7.

[17] H. Yasumatsu and T. Kondow. Reactive scattering of clusters and cluster ions from solid surfaces. *Rep. Prog. Phys.*, 66(10):1783–1832, 2003. doi: 10.1088/0034-4885/66/10/R06.

[18] L. D. Socaciu, J. Hagen, T. M. Bernhardt, L. Wöste, U. Heiz, H. Häkkinen, and U. Landman. Catalytic CO oxidation by free Au_2^-: Experiment and theory. *J. Am. Chem. Soc.*, 125(34):10437–10445, 2003. doi: 10.1021/ja027926m.

[19] J. Libuda and H.-J. Freund. Molecular beam experiments on model catalysts. *Surf. Sci. Rep.*, 57(7-8):157–298, 2005. doi: 10.1016/j.surfrep.2005.03.002.

[20] W. Gerlach and O. Stern. Der experimentelle Nachweis des magnetischen Moments des Silberatoms. *Z. f. Phys. A*, 8(1):110–111, 1922. doi: 10.1007/BF01329580.

[21] A. W. Kleyn. Molecular beams and chemical dynamics at surfaces. *Chem. Soc. Rev.*, 32(2):87–95, 2003. doi: 10.1039/b105760j.

[22] L. Pedemonte, A. Gussoni, R. Tatarek, and G. Bracco. High-resolution scattering apparatus for surface studies. *Rev. Sci. Instrum.*, 73(12):4257–4263, 2002. doi: 10.1063/1.1517147.

[23] D. P. Woodruff, editor. *Surface dynamics*, volume 11 of *The chemical physics of solid surfaces*. Elsevier Science, 1. edition, 2003. ISBN 9780444514370.

[24] H. Ibach. *Physics of Surfaces and Interfaces*. Springer, Berlin, 1 edition, 2006. ISBN 978-3-540-34709-5.

[25] T. Kondo, H. S. Kato, T. Yamada, S. Yamamoto, and M. Kawai. Effect of the molecular structure on the gas-surface scattering studied by supersonic molecular beam. *Eur. Phys. J. D*, 38(1):129–138, 2006. doi: 10.1140/epjd/e2005-00284-4.

[26] W. Christen, T. Krause, and K. Rademann. Transferring pharmaceuticals into the gas phase. *Int. J. Mass Spectrom.*, 277(1-3):305–308, 2008. doi: 10.1016/j.ijms.2008.04.029.

[27] L. Bergmann, C. Schaefer, and K. Kleinermanns, editors. *Lehrbuch der Experimentalphysik: Lehrbuch der Experimentalphysik 5. Gase, Nanosysteme, Flüssigkeiten*, volume 5. Gruyter, 2. edition, 2005. ISBN 3110174847.

[28] M. Kappes and S. Leutwyler. *Atomic and Molecular Beam Methods: Volume 1*, chapter Molecular beams of clusters. Oxford University Press, USA, illustrated edition, 1988. ISBN 0195042808. G. Scoles, D. Bassi, U. Buck and D. C. Lainé, (editors).

[29] J. Wang, V. A. Shamamian, B. R. Thomas, J. M. Wilkinson, J. Riley, C. F. Giese, and W. R. Gentry. Speed ratios greater than 1000 and temperatures less than 1 mK in a pulsed He beam. *Phys. Rev. Lett.*, 60(8):696–699, 1988. doi: 10.1103/PhysRevLett.60.696.

[30] U. Even, J. Jortner, D. Noy, N. Lavie, and C. Cossart-Magos. Cooling of large molecules below 1 K and He clusters formation. *J. Chem. Phys.*, 112(18):8068–8071, 2000. doi: 10.1063/1.481405.

[31] W. Christen, K. Rademann, and U. Even. Efficient cooling in supersonic jet expansions of supercritical fluids: CO and CO_2. *J. Chem. Phys.*, 125(17):174307-5, 2006. doi: 10.1063/1.2364505.

[32] M. A. Duncan. Spectroscopy of metal ion complexes: Gas-phase models for solvation. *Annu. Rev. Phys. Chem.*, 48:69–93, 1997. doi: 10.1146/annurev.physchem.48.1.69.

[33] V. J. Herrero and I. Tanarro. Production and applications of atomic and molecular beams. *Vacuum*, 52(1-2):3–10, 1999. doi: 10.1016/S0042-207X(98)00218-8.

[34] M. Havenith. *Infrared Spectroscopy of Molecular Clusters*, volume 176/2002 of *Springer Tracts in Modern Physics*. Springer, Berlin, 2002. ISBN 978-3-540-42691-2. doi: 10.1007/3-540-45457-8.

[35] D. H. Levy. Laser spectroscopy of cold gas-phase molecules. *Annu. Rev. Phys. Chem.*, 31:197–225, 1980. doi: 10.1146/annurev.pc.31.100180.001213.

[36] J. Koperski. Study of diatomic van der Waals complexes in supersonic beams. *Phys. Rep.*, 369(3):177–326, 2002. doi: 10.1016/S0370-1573(02)00200-4.

[37] P. Wegener. *Molecular beams and low density gasdynamics*. M. Dekker, New York, 1974. ISBN 9780824761998.

[38] J. M. Prausnitz and P. R. Benson. Effective collision diameters and correlation of some thermodynamic properties of solutions. *AIChE Journal*, 5(3):301–303, 1959. doi: 10.1002/aic.690050310.

[39] J. B. Anderson and J. B. Fenn. Velocity distributions in molecular beams from nozzle sources. *Phys. Fluids*, 8(5):780–787, 1965. doi: 10.1063/1.1761320.

[40] R. W. Barber and D. R. Emerson. *Advances in Fluid Mechanics IV*, volume 32 of *Advances in Fluid Mechanics*, chapter The influence of Knudsen number on the hydrodynamic development length within parallel plate micro-channels, pages 207–216. WIT Press, Southampton, UK, 2002. ISBN 978-1-85312-910-0. M. Rahman, R. Verhoeven, C. A. Brebbia, (editors).

[41] C. Shen. *Rarefied Gas Dynamics: Fundamentals, Simulations and Micro Flows*. Springer, Berlin, 1 edition, 2005. ISBN 354023926X.

[42] R. E. Miller. *Atomic and Molecular Beam Methods: Volume 2*, chapter Free Jet Sources. Oxford University Press, USA, illustrated edition, 1992. ISBN 0195042816. G. Scoles, D. C. Lainé and U. Valbusa, (editors).

[43] H. Haberland (ed.). *Clusters of Atoms and Molecules I*. Springer, Berlin, korr. nachdruck der 1. a. von 1994 edition, 1994. ISBN 354053332X.

[44] L. W. Bruch, W. Schöllkopf, and J. P. Toennies. The formation of dimers and trimers in free jet ^{4}He cryogenic expansions. *J. Chem. Phys.*, 117(4):1544–1566, 2002. doi: 10.1063/1.1486442.

[45] R. E. Smalley, L. W., and D. H. Levy. The fluorescence excitation spectrum of rotationally cooled NO_2. *J. Chem. Phys.*, 63(11):4977–4989, 1975. doi: 10.1063/1.431244.

[46] H. Haberland, U. Buck, and M. Tolle. Velocity distribution of supersonic nozzle beams. *Rev. Sci. Instrum.*, 56(9):1712–1716, 1985. doi: 10.1063/1.1138129.

[47] R. E. Smalley, L. Wharton, and D. H. Levy. Molecular optical spectroscopy with supersonic beams and jets. *Acc. Chem. Res.*, 10(4):139–145, 1977. doi: 10.1021/ar50112a006.

[48] A. Kantrowitz and J. Grey. A high intensity source for the molecular beam. Part I. theoretical. *Rev. Sci. Instrum.*, 22(5):328–332, 1951. doi: 10.1063/1.1745921.

[49] H. Pauly. *Atom, Molecule, and Cluster Beams I: Basic Theory, Production and Detection of Thermal Energy Beams*. Springer, Berlin, 1 edition, 2000. ISBN 3540669450.

[50] J. B. Fenn. Mass spectrometric implications of high-pressure ion sources. *Int. J. Mass Spectrom.*, 200(1-3):459–478, 2000. doi: 10.1016/S1387-3806(00)00328-6.

[51] A. Ramos, J. M. Fernández, G. Tejeda, and S. Montero. Quantitative study of cluster growth in free-jet expansions of CO_2 by Rayleigh and Raman scattering. *Phys. Rev. A*, 72(5):053204–7, 2005. doi: 10.1103/PhysRevA.72.053204.

[52] P. K. Sharma, E. L. Knuth, and W. S. Young. Species enrichment due to mach-number focusing in a molecular-beam mass-spectrometer sampling system. *J. Chem. Phys.*, 64(11):4345–4351, 1976. doi: 10.1063/1.432103.

[53] G. A. Bird. Transition regime behavior of supersonic beam skimmers. *Phys. Fluids*, 19(10):1486–1491, 1976. doi: 10.1063/1.861351.

[54] R. Campargue. Progress in overexpanded supersonic jets and skimmed molecular beams in free-jet zones of silence. *J. Phys. Chem.*, 88(20):4466–4474, 1984. doi: 10.1021/j150664a004.

[55] A. Amrein, M. Quack, and U. Schmitt. High-resolution interferometric fourier transform infrared absorption spectroscopy in supersonic free jet expansions: carbon monoxide, nitric oxide, methane, ethyne, propyne, and trifluoromethane. *J. Phys. Chem.*, 92(19):5455–5466, 1988. doi: 10.1021/j100330a025.

[56] G. M. McClelland, K. L. Saenger, J. J. Valentini, and D. R. Herschbach. Vibrational and rotational relaxation of iodine in seeded supersonic beams. *J. Phys. Chem.*, 83(8):947–959, 1979. doi: 10.1021/j100471a014.

[57] C. H. Sin, H. M. Pang, D. M. Lubman, and J. Zorn. Supercritical carbon dioxide injection in supersonic beam mass spectrometry. *Anal. Chem.*, 58(2):487–490, 1986. doi: 10.1021/ac00293a051.

[58] W. S. Young. Derivation of the free-jet mach-disk location using the entropy-balance principle. *Phys. Fluids*, 18(11):1421–1425, 1975. doi: 10.1063/1.861039.

[59] J. P. Toennies and K. Winkelmann. Theoretical studies of highly expanded free jets: Influence of quantum effects and a realistic intermolecular potential. *J. Chem. Phys.*, 66(9):3965–3979, 1977. doi: 10.1063/1.434448.

[60] L. K. Randeniya and M. A. Smith. A study of molecular supersonic flow using the generalized boltzmann equation. *J. Chem. Phys.*, 93(1):661–673, 1990. doi: 10.1063/1.459514.

[61] T. D. Gaily, S. D. Rosner, and R. A. Holt. New laser-induced fluorescence method for molecular beam velocity analysis. *Rev. Sci. Instrum.*, 47(1):143–145, 1976. doi: 10.1063/1.1134464.

[62] D. Keil, A. Lübbert, and K. Schügerl. Supercooled molecules in molecular beams of ICl and IBr. *J. Chem. Phys.*, 79(8):3845–3850, 1983. doi: 10.1063/1.446249.

[63] S. Yagyu, H. Kita, Y. Kino, K. Ozeki, M. Sasaki, and S. Yamamoto. Velocity distribution measurements of a supersonic methane molecular beam reflected from a Pt(111) surface at various reflection angles. *Jpn. J. Appl. Phys.*, 37:5772–5774, 1998. doi: 10.1143/JJAP.37.5772.

[64] W. Li, M. J. Stirniman, and S. J. Sibener. The effect of cluster formation on mass separation in binary molecular beams. *J. Chem. Phys.*, 112(7):3208–3213, 2000. doi: 10.1063/1.480904.

[65] H. C. W. Beijerinck, R. G. J. M. Moonen, and N. F. Verster. Calibration of a time-of-flight machine for molecular beam studies. *J. Phys. E: Sci. Instrum.*, 7 (1):31–36, 1974. doi: 10.1088/0022-3735/7/1/009.

[66] H. Buchenau, E. L. Knuth, J. Northby, J. P. Toennies, and C. Winkler. Mass spectra and time-of-flight distributions of helium cluster beams. *J. Chem. Phys.*, 92(11):6875–6889, 1990. doi: 10.1063/1.458275.

[67] O. F. Hagena. Cluster ion sources (invited). In *Proceedings of the Fourth International Conference on Ion Sources*, volume 63, pages 2374–2379, Bensheim (Germany), 1992. AIP. doi: 10.1063/1.1142933.

[68] E. W. Becker, K. Bier, and W. Henkes. Strahlen aus kondensierten Atomen und Molekeln im Hochvakuum. *Z. f. Phys. A*, 146(3):333–338, 1956. doi: 10.1007/BF01330428.

[69] F. Chirot, P. Labastie, S. Zamith, and J.-M. L'Hermite. Experimental determination of nucleation scaling law for small charged particles. *Phys. Rev. Lett.*, 99 (19):193401–4, 2007. doi: 10.1103/PhysRevLett.99.193401.

[70] O. F. Hagena and W. Obert. Cluster formation in expanding supersonic jets: Effect of pressure, temperature, nozzle size, and test gas. *J. Chem. Phys.*, 56(5): 1793–1802, 1972. doi: 10.1063/1.1677455.

[71] K.-M. Weitzel and J. Mähnert. The binding energies of small Ar, CO and N_2 cluster ions. *Int. J. Mass Spectrom.*, 214(2):175–212, 2002. doi: 10.1016/S1387-3806(02)00522-5.

[72] T. A. Milne, A. E. Vandegrift, and F. T. Greene. Mass-Spectrometric observations of argon clusters in nozzle beams II. the kinetics of dimer growth. *J. Chem. Phys.*, 52(3):1552–1560, 1970. doi: 10.1063/1.1673166.

[73] J. Vigué, P. Labastie, and F. Calvo. Evidence for $N^{1/3}$ dependence of the sticking cross-section of atoms on small and medium-size van der Waals clusters. *Eur. Phys. J. D*, 8(2):265–272, 2000. doi: 10.1007/s100530050035.

[74] D. Golomb, R. E. Good, A. B. Bailey, M. R. Busby, and R. Dawbarn. Dimers, clusters, and condensation in free jets. II. *J. Chem. Phys.*, 57(9):3844–3852, 1972. doi: 10.1063/1.1678854.

[75] S. De Dea, D. R. Miller, and R. E. Continetti. Cluster and solute velocity distributions in free-jet expansions of supercritical CO_2. *J. Phys. Chem. A*, 113(2): 388–398, 2009. doi: 10.1021/jp805618z.

[76] T. A. Beu, C. Steinbach, and U. Buck. Model analysis of the fragmentation of large H_2O and NH_3 clusters based on MD simulations. *Eur. Phys. J. D*, 27(3): 223–229, 2003. doi: 10.1140/epjd/e2003-00268-4.

[77] I. Napari and H. Vehkamaki. The role of dimers in evaporation of small argon clusters. *J. Chem. Phys.*, 121(2):819–822, 2004. doi: 10.1063/1.1763148.

[78] A. Khoukaz, T. Lister, C. Quentmeier, R. Santo, and C. Thomas. Systematic studies on hydrogen cluster beam production. *Eur. Phys. J. D*, 5(2):275–281, 1999. doi: 10.1007/s100500050286.

[79] T. Jiang and J. A. Northby. Fragmentation clusters formed in supercritical expansions of ^{4}He. *Phys. Rev. Lett.*, 68(17):2620–2623, 1992. doi: 10.1103/PhysRevLett.68.2620.

[80] E. L. Knuth, F. Schünemann, and J. P. Toennies. Supercooling of H_2 clusters produced in free-jet expansions from supercritical states. *J. Chem. Phys.*, 102 (15):6258–6271, 1995. doi: 10.1063/1.469072.

[81] J. Harms, J. P. Toennies, and E. L. Knuth. Droplets formed in helium free-jet expansions from states near the critical point. *J. Chem. Phys.*, 106(8):3348–3357, 1997. doi: 10.1063/1.473083.

[82] E. L. Knuth and U. Henne. Average size and size distribution of large droplets produced in a free-jet expansion of a liquid. *J. Chem. Phys.*, 110(5):2664–2668, 1999. doi: 10.1063/1.477988.

[83] E. Parra, S. J. McNaught, and H. M. Milchberg. Characterization of a cryogenic, high-pressure gas jet operated in the droplet regime. *Rev. Sci. Instrum.*, 73(2): 468–475, 2002. doi: 10.1063/1.1433945.

[84] O. F. Hagena. Condensation in free jets: Comparison of rare gases and metals. *Z. f. Phys. D*, 4(3):291–299, 1987. doi: 10.1007/BF01436638.

[85] O. F. Hagena. Silver clusters from nozzle expansions. *Z. f. Phys. D*, 17(3):157–158, 1990. doi: 10.1007/BF01437893.

[86] O. F. Hagena. Formation of silver clusters in nozzle expansions. *Z. f. Phys. D*, 20(1):425–428, 1991. doi: 10.1007/BF01544028.

[87] J. Wörmer, V. Guzielski, J. Stapelfeldt, and T. Möller. Fluorescence excitation spectroscopy of xenon clusters in the VUV. *Chem. Phys. Lett.*, 159(4):321–326, 1989. doi: 10.1016/0009-2614(89)87493-7.

[88] R. A. Smith, T. Ditmire, and J. W. G. Tisch. Characterization of a cryogenically cooled high-pressure gas jet for laser/cluster interaction experiments. *Rev. Sci. Instrum.*, 69(11):3798–3804, 1998. doi: 10.1063/1.1149181.

[89] T. Hirayama, A. Kanehira, and I. Arakawa. Supersonic cluster beam source using a differential cryopumping system. *Rev. Sci. Instrum.*, 64(4):962–965, 1993. doi: 10.1063/1.1144150.

[90] A. V. Malakhovskii. Time evolution of pulsed supersonic jets of argon clusters. *J. Phys. D*, 33(5):556–563, 2000. doi: 10.1088/0022-3727/33/5/311.

[91] J.-H. Song, S. N. Kwon, D.-K. Choi, and W.-K. Choi. Assessment of an ionized CO_2 gas cluster accelerator. *Nucl. Instr. and Meth. B*, 179(4):568–574, 2001. doi: 10.1016/S0168-583X(01)00686-3.

[92] I. V. Hertel and W. Radloff. Ultrafast dynamics in isolated molecules and molecular clusters. *Rep. Prog. Phys.*, 69(6):1897–2003, 2006. ISSN 0034-4885. doi: 10.1088/0034-4885/69/6/R06.

[93] A. V. Malakhovskii. Ejection of cluster ions as a result of electron impact ionization of argon clusters. *Chem. Phys.*, 270(3):471–481, 2001. doi: 10.1016/S0301-0104(01)00419-0.

[94] G. N. Makarov. Cluster temperature. Methods for its measurement and stabilization. *Phys.-Usp.*, 51(4):319–353, 2008. doi: 10.1070/PU2008v051n04ABEH006421.

[95] E. L. Knuth. Size correlations for condensation clusters produced in free-jet expansions. *J. Chem. Phys.*, 107(21):9125–9132, 1997. doi: 10.1063/1.475204.

[96] K. Y. Kim, V. Kumarappan, and H. M. Milchberg. Measurement of the average size and density of clusters in a gas jet. *Appl. Phys. Lett.*, 83(15):3210–3212, 2003. doi: 10.1063/1.1618017.

[97] A. Bonnamy, R. Georges, A. Benidar, J. Boissoles, A. Canosa, and B. R. Rowe. Infrared spectroscopy of $(CO_2)_N$ nanoparticles ($30 < N < 14500$) flowing in a uniform supersonic expansion. *J. Chem. Phys.*, 118(8):3612–3621, 2003. doi: 10.1063/1.1539036.

[98] U. Buck and H. Meyer. Electron bombardment fragmentation of Ar van der Waals clusters by scattering analysis. *J. Chem. Phys.*, 84(9):4854–4861, 1986. doi: 10.1063/1.449974.

[99] F. Huisken and T. Pertsch. Infrared photodissociation and cluster-specific detection of internally cold $(C_2H_4)_n$ van der Waals complexes. *J. Chem. Phys.*, 86(1): 106–113, 1987. doi: 10.1063/1.452601.

[100] Z. Herman. The crossed-beam scattering method in studies of ion-molecule reaction dynamics. *Int. J. Mass Spectrom.*, 212(1-3):413–443, 2001. doi: 10.1016/S1387-3806(01)00493-6.

[101] I. Pócsik. Lognormal distribution as the natural statistics of cluster systems. *Z. f. Phys. D*, 20(1):395–397, 1991. doi: 10.1007/BF01544020.

[102] K. J. Mendham, N. Hay, M. B. Mason, J. W. G. Tisch, and J. P. Marangos. Cluster-size distribution effects in laser-cluster interaction experiments. *Phys. Rev. A*, 64(5):055201–4, 2001. doi: 10.1103/PhysRevA.64.055201.

[103] O. Echt, K. Sattler, and E. Recknagel. Magic numbers for sphere packings: Experimental verification in free xenon clusters. *Phys. Rev. Lett.*, 47(16):1121–1124, 1981. doi: 10.1103/PhysRevLett.47.1121.

[104] E. E. Polymeropoulos and J. Brickmann. Magic numbers in ionized rare-gas clusters. *Surf. Sci.*, 156(Part 2):563–571, 1985. doi: 10.1016/0039-6028(85)90225-0.

[105] I. A. Harris, R. S. Kidwell, and J. A. Northby. Structure of charged argon clusters formed in a free jet expansion. *Phys. Rev. Lett.*, 53(25):2390–2393, 1984. doi: 10.1103/PhysRevLett.53.2390.

[106] O. Kandler, T. Leisner, O. Echt, and E. Recknagel. Carbon monoxide clusters: critical size and magic numbers. *Z. f. Phys. D*, 10(2):295–301, 1988. doi: 10.1007/BF01384864.

[107] S. R. Desai, C. S. Feigerle, and J. C. Miller. Magic numbers in $(NO)_m^+ Ar_n$ heteroclusters produced by two-photon ionization in a supersonic expansion. *J. Chem. Phys.*, 97(3):1793–1799, 1992. doi: 10.1063/1.463166.

[108] W. Greiner and A. Solov'yov. Atomic cluster physics: new challenges for theory and experiment. *Chaos, Solitons and Fractals*, 25(4):835–843, 2005. doi: 10.1016/j.chaos.2004.11.077.

[109] W. E. Stephens. A pulsed mass spectrometer with time dispersion. *Phys. Rev.*, 69(11-12):691, 1946. doi: 10.1103/PhysRev.69.674.2.

[110] A. E. Cameron and Jr. Eggers. An ion "velocitron". *Rev. Sci. Instrum.*, 19(9):605–607, 1948. doi: 10.1063/1.1741336.

[111] W. C. Wiley and I. H. McLaren. Time-of-flight mass spectrometer with improved resolution. *Rev. Sci. Instrum.*, 26(12):1150–1157, 1955. doi: 10.1063/1.1715212.

[112] B. A. Mamyrin. Time-of-flight mass spectrometry (concepts, achievements, and prospects). *Int. J. Mass Spectrom.*, 206(3):251–266, 2001. doi: 10.1016/S1387-3806(00)00392-4.

[113] U. Boesl, R. Weinkauf, and E. W. Schlag. Reflectron time-of-flight mass spectrometry and laser excitation for the analysis of neutrals, ionized molecules and secondary fragments. *Int. J. Mass Spectrom. Ion Processes*, 112(2-3):121–166, 1992. doi: 10.1016/0168-1176(92)80001-H.

[114] K. Håkansson, R. A. Zubarev, P. Håkansson, V. Laiko, and A. F. Dodonov. Design and performance of an electrospray ionization time-of-flight mass spectrometer. *Rev. Sci. Instrum.*, 71(1):36–41, 2000. doi: 10.1063/1.1150157.

[115] M. Karas, D. Bachmann, U. Bahr, and F. Hillenkamp. Matrix-assisted ultraviolet laser desorption of non-volatile compounds. *Int. J. Mass Spectrom. Ion Processes*, 78:53–68, 1987. doi: 10.1016/0168-1176(87)87041-6.

[116] F. Hillenkamp and M. Karas. Matrix-assisted laser desorption/ionisation, an experience. *Int. J. Mass Spectrom.*, 200(1-3):71–77, 2000. doi: 10.1016/S1387-3806(00)00300-6.

[117] U. Rohner, J. A. Whitby, and P. Wurz. A miniature laser ablation time-of-flight mass spectrometer for in situ planetary exploration. *Meas. Sci. Technol.*, 14(12): 2159–2164, 2003. doi: 10.1088/0957-0233/14/12/017.

[118] D. Price and G. J. Milnes. Recent developments in techniques utilising time-of-flight mass spectrometry. *Int. J. Mass Spectrom. Ion Processes*, 60(1):61–81, 1984. doi: 10.1016/0168-1176(84)80076-2.

[119] D. Price and G. J. Milnes. The renaissance of time-of-flight mass spectrometry. *Int. J. Mass Spectrom. Ion Processes*, 99(1-2):1–39, 1990. doi: 10.1016/0168-1176(90)85019-X.

[120] H. Wollnik. Time-of-flight mass analyzers. *Mass Spectrom. Rev.*, 12(2):89–114, 1993. doi: 10.1002/mas.1280120202.

[121] E. W. Schlag, editor. *Time of Flight Mass Spectrometry and Its Applications*. Elsevier Science Ltd, illustrated edition, 1994. ISBN 0444818758.

[122] C. Weickhardt, F. Moritz, and J. Grotemeyer. Time-of-flight mass spectrometry: State-of the-art in chemical analysis and molecular science. *Mass Spectrom. Rev.*, 15(3):139–162, 1996. doi: 10.1002/(SICI)1098-2787(1996)15:3<139::AID-MAS1>3.0.CO;2-J.

[123] M. Guilhaus, D. Selby, and V. Mlynski. Orthogonal acceleration time-of-flight mass spectrometry. *Mass Spectrom. Rev.*, 19(2):65–107, 2000. doi: 10.1002/(SICI)1098-2787(2000)19:2<65::AID-MAS1>3.0.CO;2-E.

[124] I. V. Chernushevich, A. V. Loboda, and B. A. Thomson. An introduction to quadrupole-time-of-flight mass spectrometry. *J. Mass Spectrom.*, 36(8):849–865, 2001. doi: 10.1002/jms.207.

[125] S. Humphries. *Principles of Charged Particle Acceleration*. Wiley-Interscience, 1986. ISBN 0471878782.

[126] D. Ioanoviciu. Ion-optical properties of time-of-flight mass spectrometers. *Int. J. Mass Spectrom.*, 206(3):211–229, 2001. doi: 10.1016/S1387-3806(00)00314-6.

[127] H. Wollnik. Ion optics in mass spectrometers. *J. Mass Spectrom.*, 34(10):991–1006, 1999. doi: 10.1002/(SICI)1096-9888(199910)34:10<991::AID-JMS870>3.0.CO;2-1.

[128] A. A. Makarov. Ideal and quasi-ideal time focusing of charged particles. *J. Phys. D*, 24(4):533–540, 1991. doi: 10.1088/0022-3727/24/4/003.

[129] M. Vestal and P. Juhasz. Resolution and mass accuracy in matrix-assisted laser desorption ionization- time-of-flight. *J. Am. Soc. Mass Spectrom.*, 9(9):892–911, 1998. doi: 10.1016/S1044-0305(98)00069-5.

[130] D. M. Lubman and R. M. Jordan. Design for improved resolution in a time-of-flight mass spectrometer using a supersonic beam and laser ionization source. *Rev. Sci. Instrum.*, 56(3):373–376, 1985. doi: 10.1063/1.1138306.

[131] J. E. Pollard and R. B. Cohen. Electron-impact ionization time-of-flight mass spectrometer for molecular beams. *Rev. Sci. Instrum.*, 58(1):32–37, 1987. doi: 10.1063/1.1139562.

[132] Y. H. Chen, M. Gonin, K. Fuhrer, A. Dodonov, C. S. Su, and H. Wollnik. Orthogonal electron impact source for a time-of-flight mass spectrometer with high mass resolving power. *Int. J. Mass Spectrom.*, 185-187:221–226, 1999. doi: 10.1016/S1387-3806(98)14152-0.

[133] F. Dubois, R. Knochenmuss, and R. Zenobi. Optimization of an ion-to-photon detector for large molecules in mass spectrometry. *Rapid Commun. Mass Spectrom.*, 13(19):1958–1967, 1999. doi: 10.1002/(SICI)1097-0231(19991015)13:19<1958::AID-RCM738>3.0.CO;2-3.

[134] M. Guilhaus, V. Mlynski, and D. Selby. Perfect timing: Time-of-flight mass spectrometry. *Rapid Commun. Mass Spectrom.*, 11(9):951–962, 1997. doi: 10.1002/(SICI)1097-0231(19970615)11:9<951::AID-RCM785>3.0.CO;2-H.

[135] D. Ioanoviciu. Ion-optical solutions in time-of-flight mass spectrometry. *Rapid Commun. Mass Spectrom.*, 9(11):985–997, 1995. doi: 10.1002/rcm.1290091104.

[136] R. Stein. Space and velocity focusing in time-of-flight mass spectrometers. *Int. J. Mass Spectrom. Ion Phys.*, 14(2):205–218, 1974. doi: 10.1016/0020-7381(74)80008-2.

[137] J. Jauhiainen, S. Aksela, and E. Nõmmiste. Flight time distribution and collection efficiency studies for time-of-flight mass spectrometer. *Phys. Scr.*, 51(5):549–556, 1995. doi: 10.1088/0031-8949/51/5/002.

[138] R. B. Opsal, K. G. Owens, and J. P. Reilly. Resolution in the linear time-of-flight mass spectrometer. *Anal. Chem.*, 57(9):1884–1889, 1985. doi: 10.1021/ac00286a020.

[139] M. Guilhaus. Special feature: Tutorial. Principles and instrumentation in time-of-flight mass spectrometry. Physical and instrumental concepts. *J. Mass Spectrom.*, 30(11):1519–1532, 1995. doi: 10.1002/jms.1190301102.

[140] M. Yang and J. P. Reilly. A reflectron mass spectrometer with UV laser-induced surface ionization. *Int. J. Mass Spectrom. Ion Processes*, 75(2):209–219, 1987. doi: 10.1016/0168-1176(87)83055-0.

[141] J. A. Browder, R. L. Miller, W. A. Thomas, and G. Sanzone. High-resolution TOF mass spectrometry. II. experimental confirmation of impulse-field focusing theory. *Int. J. Mass Spectrom. Ion Phys.*, 37(1):99–108, 1981. doi: 10.1016/0020-7381(81)80112-X.

[142] G. E. Yefchak, C. G. Enke, and J. F. Holland. Models for mass-independent space and energy focusing in time-of-flight mass spectrometry. *Int. J. Mass Spectrom. Ion Processes*, 87(3):313–330, 1989. doi: 10.1016/0168-1176(89)80031-X.

[143] V. M. Collado, C. R. Ponciano, F. A. Fernandez-Lima, and E. F. da Silveira. Analysis of ion dynamics and peak shapes for delayed extraction time-of-flight mass spectrometers. *Rev. Sci. Instrum.*, 75(6):2163–2170, 2004. doi: 10.1063/1.1711161.

[144] J. Franzen. Improved resolution for MALDI-TOF mass spectrometers: a mathematical study. *Int. J. Mass Spectrom. Ion Processes*, 164(1-2):19–34, 1997. doi: 10.1016/S0168-1176(97)00049-9.

[145] D. P. Seccombe and T. J. Reddish. Theoretical study of space focusing in linear time-of-flight mass spectrometers. *Rev. Sci. Instrum.*, 72(2):1330–1338, 2001. doi: 10.1063/1.1336824.

[146] U. Even and B. Dick. Optimization of a one-dimensional time-of-flight mass spectrometer. *Rev. Sci. Instrum.*, 71(12):4421–4430, 2000. doi: 10.1063/1.1322584.

[147] F. Chandezon, B. Huber, and C. Ristori. A new-regime Wiley-McLaren time-of-flight mass spectrometer. *Rev. Sci. Instrum.*, 65(11):3344–3353, 1994. doi: 10.1063/1.1144571.

[148] J. H. D. Eland. Second-order space focusing in two-field time-of-flight mass spectrometers. *Meas. Sci. Technol.*, 4(12):1522–1524, 1993. doi: 10.1088/0957-0233/4/12/035.

[149] P. Piseri, S. Iannotta, and P. Milani. Parameterization of a two-stage mass spectrometer performing second-order space focusing. *Int. J. Mass Spectrom. Ion Processes*, 153(1):23–28, 1996. doi: 10.1016/0168-1176(95)04350-0.

[150] G. Sanzone. Energy resolution of the conventional Time-of-Flight mass spectrometer. *Rev. Sci. Instrum.*, 41(5):741–742, 1970. doi: 10.1063/1.1684631.

[151] U. Even and B. Dick. Computer optimization for high-resolution time-of-flight mass spectrometer. *Rev. Sci. Instrum.*, 71(12):4415–4420, 2000. doi: 10.1063/1.1322583.

Bibliography

[152] Wolfram Research Inc. *Mathematica, Version 7.0*. Wolfram Research, Champaign, Illinois, 2008. URL http://www.wolfram.com/.

[153] The Math Works Inc. *MATLAB version 6.0*. The MathWorks, Natick, Massachusetts, 2000. URL http://www.mathworks.com/.

[154] R Development Core Team. *R: A Language and Environment for Statistical Computing*. R Development Core Team, Vienna, Austria, 2008. URL http://www.R-project.org. ISBN 3-900051-07-0.

[155] R. H. Byrd, P. Lu, J. Nocedal, and C. Zhu. A limited memory algorithm for bound constrained optimization. *SIAM J. Sci. Comput.*, 16(5):1190–1208, 1995. doi: 10.1137/0916069.

[156] C. Berger. Design of rotationally symmetrical electrostatic mirror for time-of-flight mass spectrometry. *J. App. Phys.*, 54(7):3699–3703, 1983. doi: 10.1063/1.332603.

[157] T.-I. Wang, C.-W. Chu, H.-M. Hung, G.-S. Kuo, and C.-C. Han. Design parameters of dual-stage ion reflectrons. *Rev. Sci. Instrum.*, 65(5):1585–1589, 1994. doi: 10.1063/1.1144896.

[158] C. A. Flory, R. C. Taber, and G. E. Yefchak. Analytic expression for the ideal one-dimensional mirror potential yielding perfect energy focusing in TOF mass spectrometry. *Int. J. Mass Spectrom. Ion Processes*, 152(2-3):177–184, 1996. doi: 10.1016/0168-1176(95)04343-8.

[159] S. Scherer, K. Altwegg, H. Balsiger, J. Fischer, A. Jäckel, A. Korth, M. Mildner, D. Piazza, H. Reme, and P. Wurz. A novel principle for an ion mirror design in time-of-flight mass spectrometry. *Int. J. Mass Spectrom.*, 251(1):73–81, 2006. doi: 10.1016/j.ijms.2006.01.025.

[160] C.-S. Su. Multiple reflection type time-of-flight mass spectrometer with two sets of parallel-plate electrostatic fields. *Int. J. Mass Spectrom. Ion Processes*, 88(1): 21–28, 1989. doi: 10.1016/0168-1176(89)80039-4.

[161] L. Sleno and D. A. Volmer. Ion activation methods for tandem mass spectrometry. *J. Mass Spectrom.*, 39(10):1091–1112, 2004. doi: 10.1002/jms.703.

[162] A. Malakhovskii. Decomposition of ionized clusters in short supersonic pulses of argon. *Phys. Chem. Chem. Phys.*, 1(18):4187–4194, 1999. doi: 10.1039/a904781f.

[163] A. J. Stace and A. K. Shukla. The reactions of CO_2 cluster ions. *Int. J. Mass Spectrom. Ion Phys.*, 36(1):119–122, 1980. doi: 10.1016/0020-7381(80)80012-X.

[164] R. G. Cooks, J. H. Beynon, R. M. Caprioli, and G. R. Lester. *Metastable ions*. Elsevier Scientific Pub. Co., Amsterdam, New York,, 1973. ISBN 0444411194.

[165] H. Haberland (ed.) and K. H. Bowen. *Clusters of Atoms and Molecules II. Solvation and chemistry of free clusters and embedded, supported and compressed clusters.* Springer, Berlin, reprint of 1. edition, 1994. ISBN 354053332X.

[166] P. C. Engelking. Determination of cluster binding energy from evaporative lifetime and average kinetic energy release: Application to $(CO_2)_n^+$ and Ar_n^+ clusters. *J. Chem. Phys.*, 87(2):936–940, 1987. doi: 10.1063/1.453248.

[167] S. Wei and A. W. Castleman, Jr. Using reflection time-of-flight mass spectrometer techniques to investigate cluster dynamics and bonding. *Int. J. Mass Spectrom. Ion Processes*, 131:233–264, 1994. doi: 10.1016/0168-1176(93)03886-Q.

[168] R. Parajuli, S. Matt, O. Echt, A. Stamatovic, P. Scheier, and T. D. Märk. Decay reactions of rare gas cluster ions: Kinetic energy release distributions and binding energies. *Eur. Phys. J. D*, 16(1):69–72, 2001. doi: 10.1007/s100530170062.

[169] A. J. Stace, P. G. Lethbridge, J. E. Upham, and C. A. Woodward. Dynamics of cluster ion fragmentation. *J. Chem. Soc., Faraday Trans.*, 86(13):2405–2409, 1990. doi: 10.1039/FT9908602405.

[170] J. R. Stairs, T. E. Dermota, E. S. Wisniewski, and A. W. Castleman, Jr. Calculation to determine the mass of daughter ions in metastable decay. *Int. J. Mass Spectrom.*, 213(1):81–89, 2002. doi: 10.1016/S1387-3806(01)00520-6.

[171] J.-M. L'Hermite, L. Marcou, F. Rabilloud, and P. Labastie. A new method to study metastable fragmentation of clusters using a reflectron time-of-flight mass spectrometer. *Rev. Sci. Instrum.*, 71(5):2033–2037, 2000. doi: 10.1063/1.1150573.

[172] I. S. Gilmore and M. P. Seah. Static SIMS: metastable decay and peak intensities. *App. Surf. Sci.*, 144-145:26–30, 1999. doi: 10.1016/S0169-4332(98)00757-0.

[173] V. Grill, J. Shen, C. Evans, and R. Graham Cooks. Collisions of ions with surfaces at chemically relevant energies: Instrumentation and phenomena. *Rev. Sci. Instrum.*, 72(8):3149–3179, 2001. doi: 10.1063/1.1382641.

[174] P. M. St. John, R. D. Beck, and R. L. Whetten. Fragmentation and reaction processes in cluster-surface collisions. *Z. f. Phys. D*, 26(1):226–228, 1993. doi: 10.1007/BF01429152.

[175] Z. Herman. Surface collisions of small cluster ions at incident energies $10-10^2$ eV. *Int. J. Mass Spectrom.*, 233(1-3):361–371, 2004. doi: 10.1016/j.ijms.2004.01.011.

[176] A. V. Solov'yov and J.-P. Connerade, editors. *Latest Advances In Atomic Cluster Collisions: Fission, Fusion, Electron, Ion And Photon Impact.* Imperial College Press, illustrated edition, 2005. ISBN 1860944957.

[177] J. Gspann and G. Krieg. Reflection of clusters of helium, hydrogen, and nitrogen as function of the reflector temperature. *J. Chem. Phys.*, 61(10):4037–4047, 1974. doi: 10.1063/1.1681697.

[178] U. Even, P. J. de Lange, H. T. Jonkman, and J. Kommandeur. Electron emission induced by cluster bombardment of metallic surfaces. *Phys. Rev. Lett.*, 56(9): 965–967, 1986. doi: 10.1103/PhysRevLett.56.965.

[179] R. J. Holland, G. Q. Xu, J. Levkoff, Jr. A. Robertson, and S. L. Bernasek. Experimental studies of the dynamics of nitrogen van der Waals cluster scattering from metal surfaces. *J. Chem. Phys.*, 88(12):7952–7963, 1988. doi: 10.1063/1.454252.

[180] P. J. De Lange, P. J. Renkema, and J. Kommandeur. Yield of exoelectrons from cluster bombardment of metallic surfaces. *J. Phys. Chem.*, 92(20):5749–5754, 1988. doi: 10.1021/j100331a040.

[181] G. Tepper and D. Miller. Diffractive scattering of hydrogen dimers from LiF(001). *Phys. Rev. Lett.*, 69(20):2927–2930, 1992. doi: 10.1103/PhysRevLett.69.2927.

[182] G. Tepper and D. Miller. Coherent scattering of the hydrogen dimer from a LiF crystal. *J. Chem. Phys.*, 98(12):9585–9594, 1993. doi: 10.1063/1.464389.

[183] P. U. Andersson, A. Tomsic, M. B. Andersson, and J. B. C. Petterson. Emission of small fragments during water cluster collisions with a graphite surface. *Chem. Phys. Lett.*, 279(1-2):100–106, 1997. doi: 10.1016/S0009-2614(97)00990-1.

[184] P. U. Andersson and J. B. C. Pettersson. Ionization of water clusters by collisions with graphite surfaces. *Z. f. Phys. D*, 41(1):57–62, 1997. doi: 10.1007/s004600050289.

[185] C. Menzel, R. Baumfalk, and H. Zacharias. Angular and velocity distributions of small cluster fragments in neutral $(NH_3)_n$ scattering off LiF(100). *Chem. Phys.*, 239(1-3):287–298, 1998. doi: 10.1016/S0301-0104(98)00321-8.

[186] U. Heiz. Size-selected, supported clusters: the interaction of carbon monoxide with nickel clusters. *Applied Physics A: Materials Science & Processing*, 67(6): 621–626, 1998. doi: 10.1007/s003390050833.

[187] M. Beutl, J. Lesnik, and K. D. Rendulic. Adsorption dynamics for CO, CO-clusters and H_2 (D_2) on rhodium (111). *Surf. Sci.*, 429(1-3):71–83, 1999. doi: 10.1016/S0039-6028(99)00340-4.

[188] A. Tomsic, P. U. Andersson, N. Markovic, and J. B. C. Pettersson. Collision dynamics of large water clusters on graphite. *J. Chem. Phys.*, 119(9):4916–4922, 2003. doi: 10.1063/1.1594717.

[189] V. V. Gridin, C. R. Gebhardt, A. Tomsic, I. Schechter, H. Schröder, and K. L. Kompa. Surface impact induced fragmentation and charging of neat and mixed clusters of SO_2 and H_2O. *Int. J. Mass Spectrom.*, 232(1):1–7, 2004. doi: 10.1016/j.ijms.2003.10.003.

[190] P. Piseri, H. V. Tafreshi, and P. Milani. Manipulation of nanoparticles in supersonic beams for the production of nanostructured materials. *Curr. Opin. Solid State Mater. Sci.*, 8(3-4):195–202, 2004. doi: 10.1016/j.cossms.2004.08.002.

[191] R. D. Beck, P. St. John, M. L. Homer, and R. L. Whetten. Impact-induced cleaving and melting of alkali-halide nanocrystals. *Science*, 253(5022):879–883, 1991. doi: 10.1126/science.253.5022.879.

[192] P. M. St. John, R. D. Beck, and R. L. Whetten. Reactions in cluster-surface collisions. *Phys. Rev. Lett.*, 69(9):1467–1470, 1992. doi: 10.1103/PhysRevLett.69.1467.

[193] P. M. St. John and R. L. Whetten. Hyperthermal collisions of silicon clusters with solid surfaces. *Chem. Phys. Lett.*, 196(3-4):330–336, 1992. doi: 10.1016/0009-2614(92)85977-I.

[194] P. M. St. John, C. Yeretzian, and R. L. Whetten. Electron emission mechanism for impact of carbon (C_N^-) and silicon (Si_N^-) clusters. *J. Phys. Chem.*, 96(23): 9100–9104, 1992. doi: 10.1021/j100202a005.

[195] S. Nonose, J. Hirokawa, M. Ichihashi, M. Sakamoto, H. Tanaka, and T. Kondow. Collision-induced reactions of size-selected molecular cluster anions. *Z. f. Phys. D*, 26(1):223–225, 1993. doi: 10.1007/BF01429151.

[196] C. Yeretzian, K. Hansen, R. D. Beck, and R. L. Whetten. Surface scattering of C_{60}^+: Recoil velocities and yield of C_{60}. *J. Chem. Phys.*, 98(9):7480–7484, 1993. doi: 10.1063/1.464687.

[197] E. Hendell, U. Even, T. Raz, and R. D. Levine. Shattering of clusters upon surface impact: An experimental and theoretical study. *Phys. Rev. Lett.*, 75(14): 2670–2673, 1995. doi: 10.1103/PhysRevLett.75.2670.

[198] R. D. Beck, P. Weis, G. Brauchle, and J. Rockenberger. Tandem time-of-flight mass spectrometer for cluster-surface scattering experiments. *Rev. Sci. Instrum.*, 66(8):4188–4197, 1995. doi: 10.1063/1.1145369.

[199] E. Hendell and U. Even. Cluster-surface interaction at high kinetic energy. I. Electron emission. *J. Chem. Phys.*, 103(20):9045–9052, 1995. doi: 10.1063/1.470015.

[200] T. Tsukuda, H. Yasumatsu, T. Sugai, A. Terasaki, T. Nagata, and T. Kondow. Collision processes of size-selected cluster anions, $(C_6F_6)_n^-$ ($n = 1-5$), with a silicon surface. *J. Phys. Chem.*, 99(17):6367–6373, 1995. doi: 10.1021/j100017a016.

[201] A. Terasaki, T. Tsukuda, H. Yasumatsu, T. Sugai, and T. Kondow. Fragmentation process of size-selected aluminum cluster anions in collision with a silicon surface. *J. Chem. Phys.*, 104(4):1387–1393, 1996. doi: 10.1063/1.470905.

[202] R. D. Beck, J. Rockenberger, P. Weis, and M. M. Kappes. Fragmentation of C_{60}^+ and higher fullerenes by surface impact. *J. Chem. Phys.*, 104(10):3638–3650, 1996. doi: 10.1063/1.471066.

[203] R. D. Beck, C. Warth, K. May, and M. M. Kappes. Surface impact induced shattering of C_{60}. detection of small C_m fragments by negative surface ionization. *Chem. Phys. Lett.*, 257(5-6):557–562, 1996. doi: 10.1016/0009-2614(96)00587-8.

[204] H. Yasumatsu, S. Koizumi, A. Terasaki, and T. Kondow. Energy redistribution in cluster-surface collision: $I_2^-(CO_2)_n$ onto silicon surface. *J. Chem. Phys.*, 105 (21):9509–9514, 1996. doi: 10.1063/1.472784.

[205] A. Terasaki, H. Yamaguchi, H. Yasumatsu, and T. Kondow. Fragmentation dynamics of silicon cluster anions in collision with a silicon surface: contrast to aluminum cluster anions. *Chem. Phys. Lett.*, 262(3-4):269–273, 1996. doi: 10.1016/0009-2614(96)01068-8.

[206] H. Yasumatsu, U. Kalmbach, S. Koizumi, A. Terasaki, and T. Kondow. Dynamic solvation effects on surface-impact dissociation of $I_2^-(CO_2)_n$. *Z. f. Phys. D*, 40 (1-4):51–54, 1997. doi: 10.1007/s004600050156.

[207] H. Yasumatsu, A. Terasaki, and T. Kondow. Splitting a chemical bond with a molecular wedge via cluster-surface collisions. *J. Chem. Phys.*, 106(9):3806–3812, 1997. doi: 10.1063/1.473434.

[208] W. Christen, U. Even, T. Raz, and R. D. Levine. The transition from recoil to shattering in cluster-surface impact: an experimental and computational study. *Int. J. Mass Spectrom. Ion Processes*, 174(1-3):35–52, 1998. doi: 10.1016/S0168-1176(97)00288-7.

[209] W. Christen, U. Even, T. Raz, and R. D. Levine. Collisional energy loss in cluster surface impact: Experimental, model, and simulation studies of some relevant factors. *J. Chem. Phys.*, 108(24):10262–73, 1998. doi: 10.1063/1.476487.

[210] W. Christen and U. Even. Cluster impact chemistry. *J. Phys. Chem. A*, 102(47): 9420–9426, 1998. doi: 10.1021/jp981874z.

[211] H. Yasumatsu, S. Koizumi, A. Terasaki, and T. Kondow. Energy redistribution in $I_2^-(CO_2)_n$ collision on silicon surface. *J. Phys. Chem. A*, 102(47):9581–9585, 1998. doi: 10.1021/jp982234z.

[212] H. Yasumatsu, A. Terasakia, and T. Kondow. Charge transfer from $I_2^-(CO_2)_n$ cluster anion to silicon surface: cluster-size dependence. *Int. J. Mass Spectrom. Ion Processes*, 174(1-3):297–303, 1998. doi: 10.1016/S0168-1176(97)00309-1.

[213] W. Christen and U. Even. Chemical reactions induced by cluster impact. *Eur. Phys. J. D*, 9(1):29–34, 1999. doi: 10.1007/s100530050394.

[214] U. Kalmbach, H. Yasumatsu, S. Koizumi, A. Terasaki, and T. Kondow. Mechanism of wedge effect in splitting of chemical bond by impact of $X_2^-(CO_2)_n$ onto silicon surface (X=Br, I). *J. Chem. Phys.*, 110(15):7443–7448, 1999. doi: 10.1063/1.478646.

[215] U. Busolt, E. Cottancin, H. Röhr, L. Socaciu, T. Leisner, and L. Wöste. Cluster-surface interaction studied by time-resolved two-photon photoemission. *Applied Physics B: Lasers and Optics*, 68(3):453–457, 1999. doi: 10.1007/s003400050648.

[216] T. Kimura, T. Sugai, and H. Shinohara. Surface-induced fragmentation of higher fullerenes and endohedral metallofullerenes. *J. Chem. Phys.*, 110(19):9681–9687, 1999. doi: 10.1063/1.478932.

[217] Y. Tai, W. Yamaguchi, Y. Maruyama, K. Yoshimura, and J. Murakami. Fragmentation and ion-scattering in the low-energy collisions of small silver cluster ions ($Ag_n^+ : n = 1-4$) with a highly oriented pyrolytic graphite surface. *J. Chem. Phys.*, 113(9):3808–3813, 2000. doi: 10.1063/1.1287658.

[218] W. Christen and U. Even. Collision-induced fragmentation and neutralization of methanol cluster cations. *Eur. Phys. J. D*, 16(1):87–90, 2001. doi: 10.1007/s100530170066.

[219] S. Koizumi, H. Yasumatsu, S. Otani, and T. Kondow. Intracluster reactions of $(CS_2)_n^-$ and $(OCS_2)_n^-$ induced by surface impact. *J. Phys. Chem. A*, 106(2): 267–271, 2002. doi: 10.1021/jp012556u.

[220] T. M. Bernhardt, B. Kaiser, and K. Rademann. Unimolecular decomposition of antimony and bismuth cluster ions studied by surface collision induced dissociation mass spectrometry. *Phys. Chem. Chem. Phys.*, 4(7):1192–1200, 2002. doi: 10.1039/b110194c.

[221] W. Christen and U. Even. Hyperthermal surface-collisions of water cluster cations. *Eur. Phys. J. D*, 24(1):283–286, 2003. doi: 10.1140/epjd/e2003-00160-3.

[222] S. Koizumi, H. Yasumatsu, S. Otani, and T. Kondow. Low-energy impact of $X^-(H_2O)_n$ (X = Cl, I) onto solid surface. *J. Chem. Phys.*, 121(10):4833–4838, 2004. doi: 10.1063/1.1778378.

[223] A. Bekkerman, A. Kaplan, E. Gordon, B. Tsipinyuk, and E. Kolodney. Above the surface multifragmentation of surface scattered fullerenes. *J. Chem. Phys.*, 120(23):11026–30, 2004. doi: 10.1063/1.1739397.

[224] S.-S. Jester, P. Weis, M. Hillenkamp, O. T. Ehrler, A. Böttcher, and M. M. Kappes. Quantifying electron transfer during hyperthermal scattering of C_{60}^+ from Au(111) and n-alkylthiol self-assembled monolayers. *J. Chem. Phys.*, 124 (14):144704–8, 2006. doi: 10.1063/1.2184309.

[225] R. Smith, D. E. Harrison, and B. J. Garrison. keV particle bombardment of semiconductors: A molecular-dynamics simulation. *Phys. Rev. B*, 40(1):93, 1989. doi: 10.1103/PhysRevB.40.93.

[226] C. L. Cleveland and U. Landman. Dynamics of cluster-surface collisions. *Science*, 257(5068):355–361, 1992. doi: 10.1126/science.257.5068.355.

[227] H.-P. Kaukonen, U. Landman, and C. L. Cleveland. Reactions in clusters. *J. Chem. Phys.*, 95(7):4997–5013, 1991. doi: 10.1063/1.461716.

[228] T. Raz and R. D. Levine. On the burning of air. *Chem. Phys. Lett.*, 246(4-5): 405–412, 1995. doi: 10.1016/0009-2614(95)01144-4.

[229] I. Schek, J. Jortner, T. Raz, and R. D. Levine. Cluster-surface impact dissociation of halogen molecules in large inert gas clusters. *Chem. Phys. Lett.*, 257(3-4):273–279, 1996. doi: 10.1016/0009-2614(96)00551-9.

[230] M. Kerford and R. P. Webb. Molecular dynamics simulation of the desorption of molecules by energetic fullerene impacts on graphite and diamond surfaces. *Nucl. Instr. and Meth. B*, 153(1-4):270–274, 1999. doi: 10.1016/S0168-583X(99)00200-1.

[231] K. Chang–Koo, A. Kubota, and D. J. Economou. Molecular dynamics simulation of silicon surface smoothing by low-energy argon cluster impact. *J. Appl. Phys.*, 86(12):6758–6762, 1999. doi: 10.1063/1.371753.

[232] H. Vach. Lost-memory model for the surface scattering of van der waals clusters. *Phys. Rev. B*, 61(3):2310, 2000. doi: 10.1103/PhysRevB.61.2310.

[233] R. Webb, M. Kerford, E. Ali, M. Dunn, L. Knowles, K. Lee, J. Mistry, and F. Whitefoot. Molecular dynamics simulation of the cluster- impact-induced molecular desorption process. *Surf. Interface Anal.*, 31(4):297–301, 2001. doi: 10.1002/sia.992.

[234] N. Chaâbane, G. Jundt, H. Vach, D. M. Koch, and G. H. Peslherbe. Cage effects and rotational hindrance in the surface scattering of large $(N_2)_n$ clusters. *Int. J. Mass Spectrom.*, 220(2):159–170, 2002. doi: 10.1016/S1387-3806(02)00690-5.

[235] T.-N. V. Nguyen, D. M. Koch, G. H. Peslherbe, and H. Vach. Molecular dissociation and vibrational excitation in the surface scattering of $(N_2)_n$ and $(O_2)_n$ clusters. *J. Chem. Phys.*, 119(14):7451–7460, 2003. doi: 10.1063/1.1597199.

[236] A. Tomsic and C. R. Gebhardt. Molecular dynamics simulations of the microsolvation of ions and molecules during cluster-surface collisions. *Chem. Phys. Lett.*, 386(1-3):55–59, 2004. doi: 10.1016/j.cplett.2004.01.023.

[237] A. Tomsic and C. R. Gebhardt. A comparative study of cluster-surface collisions: Molecular-dynamics simulations of $(H_2O)_{1000}$ and $(SO_2)_{1000}$. *J. Chem. Phys.*, 123 (6):064704–8, 2005. doi: 10.1063/1.1997109.

[238] A. Gross and R. D. Levine. Evanescent high pressure during hypersonic cluster-surface impact characterized by the virial theorem. *J. Chem. Phys.*, 123(19): 194307–11, 2005. doi: 10.1063/1.2110207.

[239] A. Kaplan, A. Bekkerman, B. Gordon, B. Tsipinyuk, M. Fleischer, and E. Kolodney. Multifragmentation in cluster-surface impact: A shattering event with a common velocity for all outgoing fragments. *Nucl. Instr. and Meth. B*, 232(1-4): 184–194, 2005. doi: 10.1016/j.nimb.2005.03.043.

[240] P. Larrégaray and Gilles H. Peslherbe. On the statistical nature of collision and surface-induced dissociation: A theoretical investigation of aluminum clusters. *J. Phys. Chem. A*, 110(4):1658–1665, 2006. doi: 10.1021/jp0544311.

[241] S. Zimmermann and H. M. Urbassek. Hyperthermal cluster-surface scattering. *Eur. Phys. J. D*, 39(3):423–432, 2006. doi: 10.1140/epjd/e2006-00118-y.

[242] S. Zimmermann and H. M. Urbassek. Nonequilibrium phenomena in N_2-cluster-surface collisions: A molecular-dynamics study of fragmentation, lateral jetting, and nonequilibrium energy distributions. *Phys. Rev. A*, 74(6):063203–8, 2006. doi: 10.1103/PhysRevA.74.063203.

[243] T. C. Germann. Large-scale molecular dynamics simulations of hyperthermal cluster impact. *Int. J. Impact Eng.*, 33(1-12):285–293, 2006. doi: 10.1016/j.ijimpeng.2006.09.049.

[244] D. C. Jacobs. The role of internal energy and approach geometry in molecule/surface reactive scattering. *J. Phys.: Condens. Matter*, 7(6):1023–1045, 1995. doi: 10.1088/0953-8984/7/6/007.

[245] A. Budrevich, B. Tsipinyuk, and E. Kolodney. Generation and energy analysis of neutral C_{60} seeded molecular beams up to 60 eV with electrostatic energy analyzer. *Chem. Phys. Lett.*, 234(1-3):253–259, 1995. doi: 10.1016/0009-2614(94)01508-S.

[246] S. Harich and A. M. Wodtke. Anisotropic translational cooling: Velocity dependence of collisional alignment in a seeded supersonic expansion. *J. Chem. Phys.*, 107(15):5983–5986, 1997. doi: 10.1063/1.474324.

[247] N. A. Smith. *Novel Approaches to Nitride Film Growth: Seeded Supersonic Molecular Beam Methods*. PhD thesis, North Carolina State University, 2003.

[248] A. Miffre, M. Jacquey, M. Büchner, G. Trénec, and J. Vigué. Anomalous cooling of the parallel velocity in seeded beams. *Phys. Rev. A*, 70(3):030701–4, 2004. doi: 10.1103/PhysRevA.70.030701.

[249] A. Miffre, M. Jacquey, M. Büchner, G. Trénec, and J. Vigué. Parallel temperatures in supersonic beams: Ultracooling of light atoms seeded in a heavier carrier gas. *J. Chem. Phys.*, 122(9):094308–10, 2005. doi: 10.1063/1.1850897.

[250] B. Gologan, J. R. Green, J. Alvarez, J. Laskin, and R. G. Cooks. Ion/surface reactions and ion soft-landing. *Phys. Chem. Chem. Phys.*, 7(7):1490–1500, 2005. doi: 10.1039/b418056a.

[251] M. Turra, B. Waldschmidt, B. Kaiser, and R. Schäfer. An improved time-of-flight method for cluster deposition and ion-scattering experiments. *Rev. Sci. Instrum.*, 79(1):013905–9, 2008. doi: 10.1063/1.2834874.

[252] R. D. Beck, P. St. John, M. M. Alvarez, F. Diederich, and R. L. Whetten. Resilience of all-carbon molecules C_{60}, C_{70}, and C_{84}: a surface-scattering time-of-flight investigation. *J. Phys. Chem.*, 95(21):8402–8409, 1991. doi: 10.1021/j100174a066.

[253] A. De Martino, M. Benslimane, M. Châtelet, F. Pradère, and H. Vach. Normal to tangential velocity conversion in cluster-surface collisions: Ar_N on graphite. *J. Chem. Phys.*, 105(17):7828–7836, 1996. doi: 10.1063/1.472563.

[254] A. Anders and G. Y. Yushkov. Measurements of secondary electrons emitted from conductive substrates under high-current metal ion bombardment. *Surface and Coatings Technology*, 136(1-3):111–116, 2001. doi: 10.1016/S0257-8972(00)01038-0.

[255] N. Winograd. The magic of cluster SIMS. *Anal. Chem.*, 77(7):142 A–149 A, 2005. doi: 10.1021/ac053355f.

[256] W. Christen and K. Rademann. Apparatus for reactive cluster-surface studies. *Rev. Sci. Instrum.*, 77(1):015109–5, 2006. doi: 10.1063/1.2162463.

[257] W. Christen, T. Krause, and K. Rademann. Precise thermodynamic control of high pressure jet expansions. *Rev. Sci. Instrum.*, 78(7):073106–3, 2007. doi: 10.1063/1.2756630.

[258] O. Kornienko, P. T. A. Reilly, W. B. Whitten, and J. M. Ramsey. Electron impact ionization in a microion trap mass spectrometer. *Rev. Sci. Instrum.*, 70(10):3907–3909, 1999. doi: 10.1063/1.1150010.

[259] H. Haberland and T. Richter. Electron attachment to HCl clusters. *Z. f. Phys. D*, 10(1):99–102, 1988. doi: 10.1007/BF01425585.

[260] C. C. Hayden, S. M. Penn, K. J. Carlson Muyskens, and F. F. Crim. A molecular beam time-of-flight mass spectrometer using low-energy-electron impact ionization. *Rev. Sci. Instrum.*, 61(2):775–782, 1990. doi: 10.1063/1.1141493.

[261] P. W. Erdman and E. C. Zipf. Low-voltage, high-current electron gun. *Rev. Sci. Instrum.*, 53(2):225–227, 1982. doi: 10.1063/1.1136932.

[262] S. Raj and D. D. Sarma. Optimization of a low-energy, high brightness electron gun for inverse photoemission spectrometers. *Rev. Sci. Instrum.*, 75(4):1020–1025, 2004. doi: 10.1063/1.1647694.

[263] J. Arol Simpson and C. E. Kuyatt. Design of low voltage electron guns. *Rev. Sci. Instrum.*, 34(3):265–268, 1963. doi: 10.1063/1.1718325.

[264] M. A. Johnson, M. L. Alexander, and W. C. Lineberger. Photodestruction cross sections for mass-selected ion clusters: $(CO_2)_n^+$. *Chem. Phys. Lett.*, 112(4):285–290, 1984. doi: 10.1016/0009-2614(84)85742-5.

[265] A. Luches, V. Nassisi, and M. R. Perrone. Movable faraday cup for high-intensity electron beam pulses. *Rev. Sci. Instrum.*, 56(5):758–759, 1985. doi: 10.1063/1.1138165.

[266] C. L. Enloe. High-resolution retarding potential analyzer. *Rev. Sci. Instrum.*, 65 (2):507–508, 1994. doi: 10.1063/1.1145167.

[267] J. Mathew, R. A. Meger, R. F. Fernsler, and J. A. Gregor. Retarding field energy analyzer for the characterization of negative glow sheet plasmas in a magnetic field. *Rev. Sci. Instrum.*, 67(8):2818–2825, 1996. doi: 10.1063/1.1147088.

[268] U. Bahr, U. Röhling, C. Lautz, K. Strupat, M. Schürenberg, and F. Hillenkamp. A charge detector for time-of-flight mass analysis of high mass ions produced by matrix-assisted laser desorption/ionization (MALDI). *Int. J. Mass Spectrom. Ion Processes*, 153(1):9–21, 1996. doi: 10.1016/0168-1176(95)04351-9.

[269] J. L. Wiza. Microchannel plate detectors. *Nucl. Instrum. Methods*, 162(1-3): 587–601, 1979. ISSN 0029-554X. doi: 10.1016/0029-554X(79)90734-1.

[270] I. S. Gilmore and M. P. Seah. Ion detection efficiency in SIMS: Dependencies on energy, mass and composition for microchannel plates used in mass spectrometry. *Int. J. Mass Spectrom.*, 202(1-3):217–229, 2000. doi: 10.1016/S1387-3806(00)00245-1.

[271] M. W. Forbes, M. Sharifi, T. Croley, Z. Lausevic, and R. E. March. Simulation of ion trajectories in a quadrupole ion trap: a comparison of three simulation programs. *J. Mass Spectrom.*, 34(12):1219–1239, 1999. doi: 10.1002/(SICI)1096-9888(199912)34:12<1219::AID-JMS897>3.0.CO;2-L.

[272] D. A. Dahl. SIMION for the personal computer in reflection. *Int. J. Mass Spectrom.*, 200(1-3):3–25, 2000. doi: 10.1016/S1387-3806(00)00305-5.

[273] D. A. Dahl. *SIMION 3D Version 7.0 User's Manual*. Idaho National Engineering and Environmental Laboratory, Idaho Falls, ID, Bechtel BWXT Idaho, LLC, 2000. URL http://simion.com/.

[274] C. R. Gebhardt. *Reactive Cluster Impact Dynamics and Ion Processes Investigated by Cluster Impact Mass Spectrometry*. PhD Thesis, Max-Planck-Institut für Quantenoptik, TU-München, 2000.

[275] A. M. Cravath. The rate of formation of negative ions by electron attachment. *Phys. Rev.*, 33(4):605, 1929.

[276] N. E. Bradbury and R. A. Nielsen. Absolute values of the electron mobility in hydrogen. *Phys. Rev.*, 49(5):388, 1936. doi: 10.1103/PhysRev.49.388.

[277] N. P. Christian, R. J. Arnold, and J. P. Reilly. Improved calibration of time-of-flight mass spectra by simplex optimization of electrostatic ion calculations. *Anal. Chem.*, 72(14):3327–3337, 2000. doi: 10.1021/ac991500h.

[278] J. Dąbek and L. Michalak. Size-dependent fragmentation of carbon dioxide clusters. *Vacuum*, 63(4):555–560, 2001. doi: 10.1016/S0042-207X(01)00239-1.

[279] E. M. Bahati, J. J. Jureta, D. S. Belic, S. Rachafi, and P. Defrance. Electron impact ionization and dissociation of CO_2^+ to C^+ and O^+. *J. Phys. B*, 34(9): 1757–1767, 2001. doi: 10.1088/0953-4075/34/9/312.

[280] N. Nishi and H. Shinohara. Intra-cluster ion-molecule reactions and photodissociation of molecular clusters. *Z. f. Phys. D*, 12(1):269–271, 1989. doi: 10.1007/BF01426952.

[281] J. de Maaijer-Gielbert, J. H. M. Beijersbergen, P. G. Kistemaker, and T. L. Weeding. Surface-induced dissociation of benzene on a PFPE liquid insulator in a time-of-flight mass spectrometer. *Int. J. Mass Spectrom. Ion Processes*, 153 (2-3):119–128, 1996. doi: 10.1016/0168-1176(96)04362-5.

[282] J. R. Stairs, K. M. Davis, and A. W. Castleman, Jr. Metastable dissociation of the zirconium Met-Car, Zr_8C_{12}, and connections to the production of the delayed atomic ion. *J. Chem. Phys.*, 117(9):4371–4375, 2002. doi: 10.1063/1.1496475.

[283] S. Wei, W. B. Tzeng, and A. W. Castleman, Jr. Kinetic energy release measurements of ammonia cluster ions during metastable decomposition and determination of cluster ion binding energies. *J. Chem. Phys.*, 92(1):332–339, 1990. doi: 10.1063/1.458434.

[284] S. A. Buzza, S. Wei, J. Purnell, and A. W. Castleman, Jr. Formation and metastable decomposition of unprotonated ammonia cluster ions upon femtosecond ionization. *J. Chem. Phys.*, 102(12):4832–4841, 1995. doi: 10.1063/1.469531.

[285] L. L. Haney and D. E. Riederer. Delayed extraction for improved resolution of ion/surface collision products by time-of-flight mass spectrometry. *Anal. Chim. Acta*, 397(1-3):225–233, 1999. doi: 10.1016/S0003-2670(99)00407-9.

[286] C. R. Gebhardt, A. Tomsic, H. Schröder, M. Dürr, and K. L. Kompa. Matrix-free formation of gas-phase biomolecular ions by soft cluster-induced desorption. *Angew. Chem. Int. Ed.*, 48(23):4162–4165, 2009. doi: 10.1002/anie.200804431.

[287] M. Châtelet, M. Benslimane, A. De Martino, F. Pradère, and H. Vach. Energy transfer during van der Waals cluster-surface collisions. *Surf. Sci.*, 352-354:50–54, 1996. doi: 10.1016/0039-6028(95)01089-0.

[288] E. E. B. Campbell, R. R. Schneider, A. Hielscher, A. Tittes, R. Ehlich, and I. V. Hertel. Collision induced fragmentation of mass-selected $(CO_2)_n^+$ clusters. *Z. f. Phys. D*, 22(2):521–527, 1992. doi: 10.1007/BF01426094.

List of Figures

2.1 Logarithmic plot of the Knudsen number Kn. 6
2.2 The schematic shape of a supersonic molecular beam expansion. . . 9
2.3 The schematic principle of a two stage Wiley-McLaren type linear TOFMS in conjunction with a potential diagram. 16
2.4 The FWHM definition of mass resolution. 17
2.5 Principle of space focusing in a three stage TOFMS. 20
2.6 Schematic principle of energy focusing with a two stage reflectron. 26
2.7 Convolution of the Maxwell-Boltzmann distribution with a step-function representing the skimmer shape. 27
2.8 Schematic diagram of metastable decay. 29
2.9 Fundamental particle-surface interaction processes observed from thermal to high collision energies. 31

3.1 Drawing of the experimental setup for cluster-surface interaction studies. 37

4.1 First order space focusing with a Wiley-McLaren type two stage accelerator calculated by numerical optimization. 43
4.2 Second order space focusing with a Wiley-McLaren type two stage accelerator calculated by numerical optimization. 44
4.3 Alternate variation of the accelerator length L_i and its influence on the space focus plane distance L_d. 45
4.4 Alternate variation of the accelerator length L_i and its influence on the scaled resolution (R/L_D). 46
4.5 Optimization with variation of the accelerator length L_1 and L_2 where L_3 is given by $L_3 = 50$ mm $-(L_1 + L_2)$. 47
4.6 Comparison of the two stage Wiley-McLaren configuration with a three stage accelerator. 48
4.7 The "ideal" accelerator configuration with $L_1 = 12$ mm, $L_2 = 12$ mm, $L_3 = 25$ mm and $\varnothing = 76$ mm. 50
4.8 A "real" accelerator configuration with $L_1 = 12$ mm, $L_2 = 12$ mm, $L_3 = 25$ mm and $\varnothing = 76$ mm with a shielding around the meshes. 51
4.9 A "real" accelerator configuration with $L_1 = 12$ mm, $L_2 = 12$ mm, $L_3 = 26.5$ mm and $\varnothing = 76$ mm with a shielding around the meshes. 52

List of Figures

4.10 A "real" accelerator configuration with $L_1 = 12$ mm, $L_2 = 12$ mm, $L_3 = 12$ mm and $\varnothing = 76$ mm with a shielding around the meshes. All three accelerator stages consist of six "pot-shaped" electrodes. 53

4.11 An acceleration stage configuration with "pot-shaped" electrodes $L_{pot} = 12$ mm, and $\varnothing = 76$ mm with a shielding around the meshes. The acceleration stage consist of two "pot-shaped" electrodes. . . 54

4.12 Behavior of the homogeneous domain simulated for different pot shaped acceleration stages. 55

4.13 Orthogonal extraction of Ar^+-ions with a three stage accelerator. 56

4.14 SIMION simulation of the Re-TOFMS setup for the orthogonal extraction of heavy cluster ions (Ar_{1000}^+) deflected by a deflector with a distance between the deflection plates of 80 mm. 59

4.15 SIMION simulation of the Re-TOFMS setup for the orthogonal extraction of heavy cluster ions (Ar_{1000}^+) deflected by a deflector with a distance between the deflection plates of 45 mm. 60

4.16 Comparison of the SIMION simulations of the two deflector geometries with different distances between the deflection plates (80 mm and 45 mm). 61

4.17 SIMION deflector simulation of the Re-TOFMS setup. Due to the shorter plates (deflection plates lengths 60 mm, and shielding plates lengths 20 mm each) beam widening at the detector "disc" is observable. 62

4.18 SIMION deflector simulation of the Re-TOFMS setup. Due to the optimized geometry all ions are transmitted by the deflector and minimal beam widening at the detector is obtained. 63

4.19 SIMION simulation of the Re-TOFMS setup for the orthogonal extraction of light cluster ions (Ar_{25}^+) with the same optimized deflector geometry as in figure (4.18) 64

4.20 Schematic views of the interleaved comb mass gate constructed for size selection of cluster ions. 65

4.21 SIMION simulation of the interleaved comb mass gate showing the decline of the potential. 66

4.22 Histogram graphs of the TOF distribution for three different cluster sizes (($CO_2)_{99}^+$, $(CO_2)_{100}^+$ and $(CO_2)_{101}^+$) recorded at the space focus plane of the TOFMS accelerator (wire plane of the mass gate). . . 67

4.23 Influence of the acceleration voltage on the resolution at the space focus plane obtained from the TOF distribution for three different cluster sizes. 68

4.24 Influence of the beam properties (beam width and beam pulse length) on the resolution at the space focus plane obtained from the TOF distribution for three different cluster sizes. 69

List of Figures

4.25 Numerical optimization results of the two stage reflectron for one dimensional ion motion. 70
4.26 SIMION simulation of the "ideal" two stage reflectron consisting of three meshes ($L_{R1} = 12$ mm, $L_{R2} = 70$ mm, $\varnothing = 120$ mm). . . 72
4.27 SIMION simulation of a "real" two stage reflectron consisting of two stages (deceleration stage: $L_{R1} = 12$ mm, reflection stage: $L_{R2} = 71$ mm, $\varnothing = 120$ mm). 73
4.28 SIMION simulation of a "real" two stage reflectron consisting of two stages (deceleration stage: $L_{R1} = 12$ mm, reflection stage: $L_{R2} = 74$ mm, $\varnothing = 120$ mm). 74
4.29 SIMION simulation of a "real" two stage reflectron consisting of two stages (deceleration stage: $L_{R1} = 12$ mm, reflection stage: $L_{R2} = 75.2$ mm, $\varnothing = 120$ mm). 75
4.30 SIMION simulation of a "real" two stage reflectron consisting of two stages (deceleration stage: $L_{R1} = 12$ mm, reflection stage: $L_{R2} = 72$ mm, $\varnothing = 120$ mm) with alternating electrode thickness geometry. 76
4.31 SIMION simulation of a "real" two stage reflectron consisting of two stages (deceleration stage: $L_{R1} = 12$ mm, reflection stage: $L_{R2} = 72.5$ mm, $\varnothing = 116$ mm)with alternating electrode thickness geometry. 77
4.32 Section view of the reflectron collider configuration. 79
4.33 TOF mass spectra of neat Argon for 3 kV acceleration and determination of the beginning cluster size. 80
4.34 Mass calibration fits for conversion of TOF spectra to mass spectra. Ar cluster sizes plotted versus TOF peak positions and fitted by two equivalent calibration functions. 81
4.35 The same TOF spectra calibrated with two different mass calibration fits and the isotopic effect. 82
4.36 Mass resolution of the two stage linear TOFMS configuration operated at 3 kV extraction. 83
4.37 Mass resolution of the two stage reflectron TOFMS configuration operated at 4 kV extraction. 84
4.38 Mass selection of a big $(CO_2)_n^+$ cluster ion with $N = 190$. 86
4.39 Depicted is the bimodal character of the cluster size distribution of $(CO)_{16}^+ - (CO)_{48}^+$ cluster ions in dependence of the valve temperature. 87
4.40 Mass spectra of $(CO_2)^+ - (CO_2)_{16}^+$ clusters in dependence of valve to e-gun distance. 88
4.41 Mass spectra of $(CO)_2^+ - (CO)_{19}^+$ clusters in dependence of valve to e-gun distance. 89
4.42 Mass spectra of $(CO_2)^+ - (CO_2)_{30}^+$ clusters in dependence of the extraction delay between valve opening and TOFMS acceleration voltage pulse. 90

173

List of Figures

4.43 Mass spectra of $(CO)_4^+ - (CO)_{99}^+$ clusters in dependence of the extraction delay between valve opening and TOFMS acceleration voltage pulse. 91

4.44 Depicted are the mass spectra for cluster sizes of $(CO)_5^+ - (CO)_{29}^+$. Comparison of the the valve mounted e-gun with the flange mounted e-gun for different ionization potentials. 92

4.45 Depicted are the mass spectra of $(CO)_2^+ - (CO)_{30}^+$ clusters for different electron gun filament currents. 94

4.46 Depicted is the pulsed operation of the flange mounted e-gun. The mass spectra show $(CO)_2^+ - (CO)_{19}^+$ clusters in dependence of the pulse duration. 95

4.47 The figure shows the unfiltered mass spectra for $(CO)_3^+ - (CO)_{28}^+$ clusters in dependence of the deflection plates potentials and the corresponding mass spectra of the size selected $(CO)_{10}^+$ cluster. 96

4.48 Collision of the CO_2^+ monomer with the stainless steel backplane of the reflectron collider. 99

4.49 Metastable decay in the different regions of the TOFMS apparatus. 101

4.50 Collision of the dimer $(CO_2)_2^+$ with the stainless steel backplane of the reflectron collider. 102

4.51 Collision of the trimer $(CO_2)_3^+$ with the stainless steel backplane of the reflectron collider. 105

4.52 Collision of the trimer $(CO_2)_3^+$ with the stainless steel backplane of the reflectron collider. 106

4.53 Collision of the tetramer $(CO_2)_4^+$ and pentamer $(CO_2)_5^+$ with the stainless steel backplane of the reflectron collider. 107

4.54 Fragment ion yield of the collision of the tetramer $(CO_2)_4^+$ and pentamer $(CO_2)_5^+$ with the stainless steel backplane of the reflectron collider. 109

4.55 Collision of the hexamer $(CO_2)_6^+$ and heptamer $(CO_2)_7^+$ with the stainless steel backplane of the reflectron collider. 110

4.56 Integrated fragment yield for the collision of the hexamer $(CO_2)_6^+$ and heptamer $(CO_2)_7^+$ with the stainless steel backplane of the reflectron collider. 112

4.57 Collision of the dimer $(CO)_2^+$ with the stainless steel backplane of the reflectron collider. 114

4.58 Collision of the trimer $(CO)_3^+$ with the stainless steel backplane of the reflectron collider. 115

4.59 Collision of the tetramer $(CO)_4^+$ with the stainless steel backplane of the reflectron collider. 116

4.60 Collision of the heptamer $(CO)_7^+$ with the stainless steel backplane of the reflectron collider. 117

List of Figures

4.61 Collision of the nonamer $(CO)_9^+$ with the stainless steel backplane of the reflectron collider. 119

4.62 Collision of the decamer $(CO)_{10}^+$ with the stainless steel backplane of the reflectron collider. 120

4.63 Collision of the $(CO)_{25}^+$ parent cluster ion with the stainless steel backplane of the reflectron collider. 122

4.64 Collision of the $(CO)_{26}^+$ parent cluster ion with the stainless steel backplane of the reflectron collider. 124

4.65 Collision of the $(CO)_{27}^+$ parent cluster ion with the stainless steel backplane of the reflectron collider. 125

4.66 Collision of the $(CO)_{30}^+$ parent cluster ion with the stainless steel backplane of the reflectron collider. 127

4.67 Collision of the $(CO)_5^+$ parent cluster ion with the SiO_2 covered silicon surface. 131

4.68 Mass spectra for two different collision energies of the $(CO)_5^+$ parent cluster ion with the SiO_2 covered silicon surface. 133

4.69 Mass spectra of the $(CO)_{10}^+$ parent cluster ion impact with the SiO_2 covered silicon surface for two different retarding field potentials U_{EF}. 134

4.70 Collision of the $(CO)_6^+$ parent cluster ion with the SiO_2 covered silicon surface. 135

4.71 Collision of the $(CO)_8^+$ parent cluster ion with the SiO_2 covered silicon surface. 137

4.72 Integrated ion yields of the impact induced products for the surface collision of the $(CO)_5^+$–$(CO)_{10}^+$ parent cluster ions with the SiO_2 covered silicon surface. 138

4.73 Integrated ion yields of the impact induced products for the surface collision of the $(CO)_{11}^+$–$(CO)_{35}^+$ parent cluster ions with the SiO_2 covered silicon surface. 140

4.74 Cluster size dependent threshold impact energy E_i for the disappearance of the impact induced fragmentation ion yields for the surface collision of the $(CO)_5^+$–$(CO)_{35}^+$ parent cluster ions with the SiO_2 covered silicon surface. 141

A.1 Pictures of the TOFMS accelerator. 176
A.2 Pictures of the TOFMS deflector. 177
A.3 Pictures of the TOFMS mass gate. 177
A.4 Pictures of the TOFMS reflectron. 178
A.5 Pictures of the retarding field energy analyzer. 178

B.1 Pictures of the nozzle mounted e-gun. 184
B.2 Pictures of the flange mounted e-gun. 186

Appendix A
TOFMS

A.1 Pictures

A.1.1 TOFMS Accelerator

Figure A.1 Shown are pictures of the TOFMS accelerator. **a)** Depicted is the TOFMS accelerator in three stage configuration without shielding. The three acceleration stages consist of six pot shaped electrodes. Each acceleration stage has a length of $L_3 = 12$ mm (1 mm spacing and 5.5 mm thickness). For the SIMION simulation of this accelerator configuration see figure (4.10). **b)** Photograph of the third acceleration stage which consists of 26 ring electrodes (0.5 mm thickness and 0.5 mm spacing). For the SIMION simulation of the accelerator configuration with the third stage consisting of this stacked ring electrode system see figure (4.9). This third stage ($L_3 = 26.5$ mm acceleration length long) was replaced by two pot shaped electrodes as shown in a.

A.1.2 TOFMS Deflector

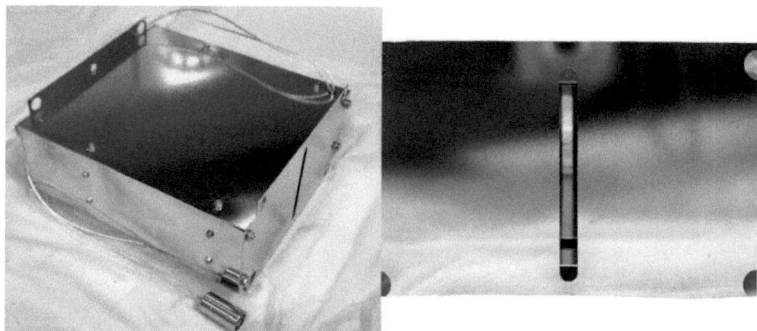

Figure A.2 Shown are pictures of the TOFMS deflector. **Left)** Depicted is the TOFMS deflector shielding enclosure. One shielded voltage cable with shielded pin connector is visible, too. **Right)** Front view of the TOFMS deflector which shows the ion entrance slit. One of the deflection plates is visible through the slit.

A.1.3 TOFMS Mass Gate

Figure A.3 Shown are pictures of the TOFMS mass gate. **Left)** Depicted is the TOFMS mass gate (shielding removed from the top and sides). Two stainless steel wires are stretched in a stainless steel frame (similar to a weaving loom). **Right)** Depicted is the side view photograph of the mass gate.

A.1.4 TOFMS Reflectron

Figure A.4 Shown are pictures of the TOFMS reflectron. **Left)** The picture shows the side view of the TOFMS reflectron (without sheet metal shielding around the reflectron body). **Right)** Photograph of the vacuum chamber with the reflectron mounted in its place. The silicon surface (dark blue) is visible through the slit of the protection sheet metal in front of the reflectron.

A.1.5 Retarding Field Energy Analyzer

Figure A.5 Shown are pictures of the rotatable retarding field energy analyzer mounted in front of the MCP-detector. **Left)** Depicted is the rotatable retarding field energy analyzer positioned in the ion beam line for energy analysis (without the MCP-detector). **Right)** Photograph of the rotatable retarding field energy analyzer in a position removed from the ion beam line for higher transmission (without the MCP-detector).

A.2 SIMION Geometry Files

A.2.1 Three Stage TOFMS Accelerator

Listing A.1 Geometry file of the three stage TOFMS accelerator

```
; Stage lengths: L1 = 12mm, L2 = 12mm, L3 = 26.5mm,
; 1. and 2. stage: electrodes with pot shape,
; 3. stage: ring electrodes
; PA definition (0.1 mm per PA-Point).
;--------------------------------------------
PA_Define(675,595,1,cylindric,y-mirrored)
;--------------------------------------------
; electrode definition
;--------------------------------------------
electrode(1) {
fill{ within {box(110,500,110,0)}
within {box(110,500,165,380)}}
}
electrode(2) {
fill{ within {box(230,500,230,0)}
within {box(175,500,285,380)}}
}
electrode(3) {
fill{ within {box(295,500,350,380)}
within {box(350,500,350,0)}}
}
electrode(4) {
fill{ within {box(355,490,360,380)}}
}
electrode(5) {
fill{    within {box(365,490,370,380)}}
}
electrode(6) {
fill{    within {box(375,490,380,380)}}
}
electrode(7) {
fill{ within {box(385,490,390,380)}}
}
electrode(8) {
fill{ within {box(395,490,400,380)}}
}
electrode(9) {
fill{ within {box(405,490,410,380)}}
}
electrode(10) {
```

```
fill{    within {box(415,490,420,380)}}
}
electrode(11) {
fill{ within {box(425,490,430,380)}}
}
electrode(12) {
fill{    within {box(435,490,440,380)}}
}
electrode(13) {
fill{ within {box(445,490,450,380)}}
}
electrode(14) {
fill{ within {box(455,490,460,380)}}
}
electrode(15) {
fill{    within {box(465,490,470,380)}}
}
electrode(16) {
fill{    within {box(475,490,480,380)}}
}
electrode(17) {
fill{ within {box(485,490,490,380)}}
}
electrode(18) {
fill{ within {box(495,490,500,380)}}
}
electrode(19) {
fill{    within {box(505,490,510,380)}}
}
electrode(20) {
fill{    within {box(515,490,520,380)}}
}
electrode(21) {
fill{    within {box(525,490,530,380)}}
}
electrode(22) {
fill{    within {box(535,490,540,380)}}
}
electrode(23) {
fill{    within {box(545,490,550,380)}}
}
electrode(24) {
fill{    within {box(555,490,560,380)}}
}
electrode(25) {
```

```
fill{    within {box(565,490,570,380)}}
}
electrode(26) {
fill{    within {box(575,490,580,380)}}
}
electrode(27) {
fill{    within {box(585,490,590,380)}}
}
electrode(28) {
fill{    within {box(595,490,600,380)}}
}
electrode(29) {
fill{    within {box(605,490,610,380)}}
}
electrode(30) {
fill{    within {box(615,490,615,0)}
within {box(615,490,620,380)}
within {box(110,595,150,590)}
within {box(190,595,655,590)}}
}
electrode(30) {
fill{    within {box(60,520,155,595)}
within {box(185,520,285,595)}
within {box(620,250,675,595)}
within {box(5,250,60,595)}
within {box(3,0,5,300)}}}
```

A.2.2 TOFMS Deflector

Listing A.2   Geometry file of the TOFMS deflector
```
; Distance between the plates: 50 mm,
; 1. shielding space: 20 mm (free space),
; 2. shielding space: 50 mm (free space),
; length deflection plates: 85 mm,
; width deflection plates: 120 mm,
; PA definition (0.5 mm per PA-Point):
;-------------------------------------------
Pa_define(109, 142, 335, planar, y-mirrored)
;-------------------------------------------
; electrode definition:
;-------------------------------------------
electrode(1){
fill{ within{box3d(3,0,60, 4,120,230)}}
}
electrode(2){
```

```
fill{ within{box3d(104,0,60,  105,120,230)}}
}
electrode(3){
fill{ within{box3d(0,0,17,  1,141,333)}
within{box3d(0,140,17,  108,141,333)}
within{box3d(107,0,17,  108,141,333)}}
}
electrode(3){
fill{ within{box3d(0,0,17,  108,141,18)}
notin{box3d(3,0,17,  105,2,18)}
within{box3d(0,0,332,  108,141,333)}
notin{box3d(3,0,332,  105,2,333)}}}
```

A.2.3 TOFMS Reflectron

Listing A.3 Geometry file of the TOFMS reflectron
```
; Stage lengths: LR1 = 12mm, LR2 = 72.5mm,
; 3mm shielding plates
; PA definition (0.1 mm per PA-Point).
;---------------------------------------------
PA_Define(960,1000,1,cylindric,y-mirrored)
;---------------------------------------------
; electrode definition
;---------------------------------------------
electrode(1) {
fill{   within {box(0,580,30,970)}}
}
electrode(1) {
fill{   within {box(30,580,85,630)}
within {box(30,630,75,700)}
within {box(30,580,30,0)}}
}
electrode(1) {
fill{   within {box(925,580,955,970)}
within {box(0,970,955,972)}}
}
electrode(2) {
fill{   within {box(150,580,150,0)}
within {box(95,580,150,630)}
within {box(105,630,150,700)}}
}
electrode(2) {
fill{ within {box(150,580,215,630)}
within {box(150,630,205,700)}}
}
```

A.2 SIMION Geometry Files

```
electrode(3) {
fill{    within {box(220,580,225,750)}}
}
electrode(4) {
fill{    within {box(230,580,360,630)}
within {box(240,630,350,700)}}
}
electrode(5) {
fill{ within {box(365,580,370,750)}}
}
electrode(6) {
fill{ within {box(375,580,505,630)}
within {box(385,630,495,700)}}
}
electrode(7) {
fill{    within {box(510,580,515,750)}}
}
electrode(8) {
fill{    within {box(520,580,650,630)}
within {box(530,630,640,700)}}
}
electrode(9) {
fill{ within {box(655,580,660,750)}}
}
electrode(10) {
fill{ within {box(665,580,795,630)}
within {box(675,630,785,700)}}
}
electrode(11) {
fill{ within {box(800,580,805,750)}}
}
electrode(12) {
fill{ within {box(810,580,875,630)}
within {box(820,630,875,700)}
within {box(875,0,875,580)}}}
```

Appendix B

Electron-Guns

B.1 Nozzle Mounted Electron-Gun

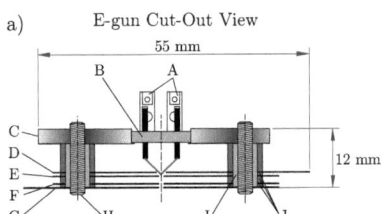

Figure B.1 Shown are pictures of the nozzle mounted e-gun. **a)** Depicted is the cut-out view of the nozzle mounted e-gun (drawing to scale): (A) cathode pin connectors, (B) thermionic emitter cathode with a tungsten filament model A 054. (C) stainless steel cathode support. (D) Wehnelt-electrode, (E) aperture-electrode, (F) extraction-electrode for pulsed electron extraction, (G) sheet metal shielding (ground potential), (H) M3 stainless steel rod, (I) alumina ceramic insulation and (J) alumina ceramic spacers. **b)** Photograph of the temperature-controlled pulsed valve with the e-gun mounted on the support of the valve body.

The nozzle mounted e-gun is based on the design of the e-gun developed by Stefan Kaesdorf[1]. The whole e-gun is nearly 25 mm in height and 50 mm in width and depth. A thermionic emitter cathode with a tungsten filament (model A 054 for the electron guns of AEI and Cambridge scanning electron microscopes) is used as electron-source. The A 054-cathode is mounted in a cathode-support with 3 mm thickness machined from non-magnetic stainless steel. The ceramic body of the cathode fits very tight in the hole of the cathode support. The other parts (electrodes) of the e-gun are also mounted on the cathode-support. Therefore four taped holes in circular arrangement for M3-rods are existent. This allows a precise arrangement of the electron optics mounted on these M3-rods. E-gun electrodes are machined from high precision stainless steel sheets with 0.15 mm

---

[1]Stefan Kaesdorf, Geraete fuer Forschung und Industrie, http://www.kaesdorf.de/

thickness (Record Metall-Folien GmbH, Germany). The electrodes are circular in shape with point welded wires on one side for the electric contact. The distance between the electrodes is defined by alumina-ceramics (99.7% Al_2O_3). The ceramic spacers are suited on a second thinner ceramic-tube which insulates the M3 screws from the electrodes. The Wehnelt-electrode is placed so that the tip of the cathode is planar with the Wehnelt-electrode (orifice $\varnothing = 1.5$ mm, 5.9 mm ceramic spacers between filament-support and Wehnelt-electrode). This arrangement allows the extraction of a high electron current from the cathode in cost of lifetime. The Wehnelt-electrode is followed by a first aperture-electrode (orifice $\varnothing = 1.6$ mm, 1 mm ceramic spacers between Wehnelt-electrode and first aperture-electrode) which is hold at negative potential to collimate the emitted electron current. An anode-electrode (orifice $\varnothing = 1.2$ mm) hold at high positive voltage (e. g. +300 V) extracts the electron current (pulsed operation possible, 1.5 mm ceramic spacers between the first aperture-electrode and "pulse" electrode). A last electrode which is hold at ground potential shields the environment from the electrode potentials (1 mm ceramic spacers between the "pulse" electrode and the shielding). Additionally a housing build up of stainless steel sheets exists which can be put on the e-gun for proper shielding. The performance of the electron gun was optimized by trying different electrode spacings and orifice diameter configurations (nearly 40 different configurations were tested). The optimization allowed improving the (Ar) cluster ion signal by a factor of three.

B.2 Flange Mounted Electron-Gun

The basic design of the flange mounted e-gun is comparable with the design of the nozzle mounted e-gun. As in the case of the nozzle mounted e-gun the filament (model A 054) is mounted in a cathode support which holds the whole assembly with (three) M3-rods. Main difference between the flange mounted e-gun and the nozzle mounted e-gun is the different electrode configuration. For the nozzle mounted e-gun sheet metal electrodes with 0.15 mm thickness were used (reduced height). Compared to this the flange mounted e-gun consists of electrodes with cylindrical shape. The design of the flange mounted e-gun is based on the designs published by Zipf et al. [261]. The Wehnelt-electrode consists of two parts. A filament tip hole electrode (orifice $\varnothing = 1.7$ mm, 5.9 mm ceramic spacers between filament-support and Wehnelt-electrode) is machined from stainless steel sheet metal with 0.15 mm thickness and a cylindrical part with 4 mm thickness ($\varnothing = 10$ mm). The other electrodes have cylindrical shapes, too ($\varnothing = 10$ mm). The first electrode is hold at negative potential and has a thickness of 9 mm. The second electrode is hold at positive potential and has a thickness of 4 mm. The last electrode is used for shielding and has a thickness of 7.5 mm (always hold at ground potential). A nickel wire mesh is mounted on the last shielding electrode hole ($\varnothing = 10$ mm) to shield the electrostatic potentials of the e-gun.

a) E-gun Cut-Out View

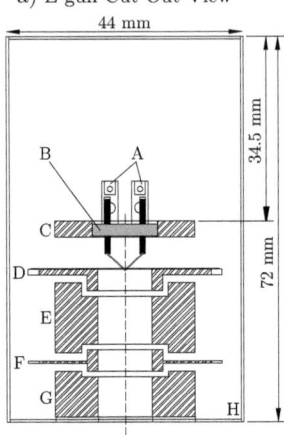

Figure B.2 Shown are pictures of the flange mounted e-gun. **a)** Depicted is the cut-out view of the flange mounted e-gun (drawing to scale): (A) cathode pin connectors, (B) thermionic emitter cathode with a tungsten filament model A 054, (C) stainless steel cathode support, (D) cylindric Wehnelt-electrode, (E) first-electrode, (F) second-electrode for electron extraction, (G) cylindric shielding-electrode (at ground potential, with wire mesh for shielding on the bottom) and (H) sheet metal shielding enclosure. Not shown are M3 support rods, alumina ceramic insulation and 1 mm alumina ceramic spacers. **b)** Photograph of the flange mounted e-gun with open housing.

All electrodes have ceramic spacers with 1 mm thickness between each other for electric insulation. Additionally metal sheet plates forming an enclosure are used to shield every side of the e-gun.

Die VDM Verlagsservicegesellschaft sucht für wissenschaftliche Verlage abgeschlossene und herausragende

Dissertationen, Habilitationen, Diplomarbeiten, Master Theses, Magisterarbeiten usw.

für die kostenlose Publikation als Fachbuch.

Sie verfügen über eine Arbeit, die hohen inhaltlichen und formalen Ansprüchen genügt, und haben Interesse an einer honorarvergüteten Publikation?

Dann senden Sie bitte erste Informationen über sich und Ihre Arbeit per Email an *info@vdm-vsg.de*.

Sie erhalten kurzfristig unser Feedback!

VDM Verlagsservicegesellschaft mbH
Dudweiler Landstr. 99
D - 66123 Saarbrücken

Telefon +49 681 3720 174
Fax +49 681 3720 1749

www.vdm-vsg.de

Die VDM Verlagsservicegesellschaft mbH vertritt

Printed by Books on Demand GmbH, Norderstedt / Germany